HOME DECORATING USING LIGHT

Golden Hands Books

Marshall Cavendish
London and New York

Edited by Gabrielle Weaver

Published by Marshall Cavendish Publications Limited,
58 Old Compton Street,
London W1V 5PA

This material was first published
by Marshall Cavendish Limited
in *Golden Homes*

This volume first published 1975

Printed in Great Britain by the Severn Valley Press Limited

ISBN 0 85685 100 0

Introduction

Creating a new and exciting environment for you and your family can be fun. But if you have experienced that moment of disappointment when despite carefully chosen colour schemes, fabrics and furniture, the overall effect hardly seems worth the effort and expense, this book has been designed with you in mind.

Planned lighting is an essential part of any home decorating scheme and yet it is so often neglected. Now you can choose the right kind of lighting for your home, to highlight or disguise, create subtle shadows or illusions of space or depth. This book will show you how to combine the special qualities of natural light, through interior glazing and the magic of mirrors, with the stunning effects of concealed lighting, spotlights and illuminated ceilings, plus specially designed plans for kitchens and hallways where efficiency and safety are at a premium.

So whether you are re-decorating or just re-vamping your home, you will find dazzling ideas for clever ways with light, to make your home a safer, more efficient and more beautiful place to live in.

RICHARD EINZIG

Contents

__Above.__ Imaginative lighting in a kitchen/ dining room, where efficiency plus style are at a premium and lighting without glare is necessary for comfortable working. The angle poise lamp and low-level table light achieve this purpose.

THE ELECTRICITY COUNCIL

Planned lighting to enhance your home

Your eyes are sensitive and delicate instruments. They scan your surroundings more efficiently than radar to provide an enormous amount of information. But they are totally dependent on one thing. Light. The reflection of light enables the eye to interpret distance, size, colour, shape, texture and movement—and the quality of this light is all-important. Not only can bad lighting spoil the appearance of your home (under a flat light velvet can look as dull as the cheapest denim) but it can also make it inefficient,—and dangerous.

To plan your lighting you must consider exactly what you need it for. Are you an amateur painter, a cook, or a carpenter? You will need suitable lighting for each hobby or activity. Is your living room open plan where the family wants to knit, do homework, and doze all at the same time? Decide what has to be lit—a work surface, a draining board, a tabletop—then how it should best be lit.

The lighting in your home deserves very careful consideration. And yet, examples of bad lighting are all too common. Often, the only source of artificial light in a living room, for example, is the single, centre light fitting. Numerous, different activities are carried out daily under this one, pitifully inadequate light. Even reading may be difficult, if you find you are sitting in your own shadow, with the light throwing distracting reflections off mirrors or other shiny surfaces.

Glare, high contrast, under-lighting, working in your own shadow—all are examples of bad lighting which can cause headaches, eyestrain, and inconvenience, and may even cause accidents.

Lighting for safety

Safety should be of first importance when planning the lighting in your home. Potential accident spots must be clearly lit at all times, especially if there are very young, elderly or short-sighted members in the family. Stairs usually present the main hazard. All stairs whether a full flight up to another floor or a couple of steps between levels, should be lit so that the beginning of the flight, and the nosings of the treads are quite obvious. Although you may know about those two unexpected steps down to the kitchen, a visitor could trip on them and fall.

Staircases are best lit from the tops of the flights so that the rises are in shadow and the nosings are accentuated. To avoid too deep a shadow on the rises, however, an infill light below the flight is also recommended. And it may be necessary to keep stairs and halls lit artificially during the day, as they often have inadequate natural light sources.

Few people consider the necessity for special night-time lighting, but it can be important. A child may have a nightmare and need reassuring; you may have to go to the bathroom; the telephone or front door bell may ring. Many emergencies can occur in the middle of the night.

Switching on lights at their normal level of intensity in a blackened room can cause temporary blindness—enough for someone to misjudge the top of the stairs. So, at night, a much lower lighting level is required for safe movement. The staircase, corridors, bathroom and telephone area can be lit effectively with 15 watt night lights and at this wattage, the cost of running the lights all night would be negligible. Regular night risers—people on shift work, or mothers with young babies to feed, for example—might also find an additional low wattage switched light useful, that they can turn on by the bed.

The hall

A hall can often benefit greatly from careful lighting as it is so often a dull, long and narrow corridor with no attractive feature. The first impression of your home needs to be bright and welcoming, and so have plenty of light here. A bright ceiling fitted downlighter is effective and downcast shadows will not matter so much in this part of the house. Otherwise try a cluster of beautiful glittering glass lamps, or a glowing ceiling lit with concealed indirect uplighters combined with attractive wall bracket lights. As halls are for 'passing through', you could try a more daring and experimental approach than would be appropriate in the living room. Just be careful not to have a brilliant lamp glaring at eye level.

The kitchen and bathroom

The kitchen is one room in the house where emphasis must be on *efficient* lighting. The work surfaces must be lit to avoid shadow and glare—but this does not mean it should be twice as bright as anywhere else in the house. The point is proficiency, not brute force. A single, central pendant is not the answer. If possible, light fittings should be directly above the work surfaces but shielded from direct view.

Strip fluorescent lighting, positioned out of sight underneath wall mounted cupboards, will shine directly onto work surfaces giving a good clear light. And fluorescent ceiling fittings are ideal for general lighting in the kitchen area. You might also consider fitting small lights inside deep storage cupboards.

Carefully placed spotlights, or tracking lights fixed to an overhead rail are also very efficient and provide lighting without glare in the kitchen. These tracks enable special light fittings to be clipped into any position on the track and can be angled at stove, sink, table, and work surfaces. (Lighting tracks, in fact, can be very useful throughout the house—as supplementary lighting in a study or workroom for example.)

Generally, there is little call for flexibility of lighting in the kitchen. But if chores such as the ironing are done there, then the tracking or spot lights could be swung into the appropriate positions. Open plan kitchen/living rooms may need a lower, more subtle, level of illumination at times, for a cosy supper or to flatter a sophisticated dinner party.

If you entertain only occasionally, then a portable light fitting that gives a soft, coloured light in the dining area can be effective. Or a pendant light, which is height-adjustable, will light the dining table effectively for both formal and informal occasions. But again, without supplementary lighting—picture lights, wall lights, or even candles—there will be a considerable, glaring contrast between the background and light fitting which can be tiring. To eliminate this, and help create a cosier atmosphere you could use a dimmer to lower the wattage of your single light.

The bathroom mainly needs a good, general purpose, overhead light, combined with mirror lights. A fixed bowl shade will not be affected too much by steam. And for very small bathrooms there are combined light/heat fittings which are useful.

Lighting the bedroom

A centre light is generally not adequate on its own in a bedroom and most people have bedside light fitments. The traditional, shaded light, however, can cast shadow and spread a diffuse light over the whole room. Although the effect may be attractive, many people enjoy reading in bed and it can be infuriating if your partner is reading when you want to sleep and the light disturbs you. This can be prevented by fitting low voltage spot lights at each side of the bed. The light will be bright enough for reading, can be angled in a suitable direction, and will leave the rest of the room in deep shadow. A soft, hidden light behind a bed alcove will add a glamorous touch.

Mirrors, particularly make-up mirrors, always need to be lit clearly. Lights positioned at either side of the glass will fall directly on the user and avoid too much glare. Or a strip light along the top is an adequate alternative; there are some effective ones available on the market which incorporate electric shaver sockets.

Living areas

The living room and dining room are likely to call for flexibility of approach. The lighting requirements of a family breakfast will be very different from those of a sophisticated dinner party in the same dining room. And you may want the option of lighting only certain areas of an open plan room, so that you can 'lose' the kitchen while dining, or simply highlight the areas you need.

The variety of effects you will need can best be produced by using several different lighting methods, each of which has a wide range of practical and decorative uses.

With *direct lighting*, such as spotlights, 90% of the light falls downwards and, although it can be very harsh, it gives a strong clear light where needed. This is ideal for reading or paper work. If the fittings are imaginatively positioned, this strong light will also bring out the form and texture of objects illuminated, and can be effective when directed onto a special ornament or other room feature. *Semi-direct lighting* also falls mainly downwards but gives a softer effect as some of the light diffuses around it.

General-diffused lighting is equally distributed upwards and downwards. This is the effect you get from most glass fitments, and round paper lampshades for example. The light is usually quite bright and generally useful, but it usually needs to be supplemented for activities that need strong light and dimmed for intimate atmosphere.

Semi-indirect lighting has to be used with care. It throws a small amount of light downwards and tends to give a rather dull, flat effect. (And yet, with a pendant fitting, this is the most commonly used lighting in the average living room.) In attractive wall or movable fitments, however, semi-indirect lighting can give a gently relaxing background light.

Ceilings would seem to benefit most from *indirect lighting* as it gives more than 90% upward light, but it does produce a shadowless, completely dull appearance. Depending on colour and intensity, it could vary from being bright and practical in the kitchen, to being soft and warm in another room. Otherwise indirect uplighters can add a touch of glamour to a wall of long curtains as long as they are positioned safely away from the fabric.

Opposite top left. *Clever use of shade and light. The spotlights throw light onto the pictures and ornaments.*
Top right. *The moveable spotlights in this black and white room provide soft light making for subtle shadows.*
Bottom. *Planned lighting in a distinctive manner, making clever use of natural light and spotlights on the ceiling.*

ELECTRICITY COUNCIL

COUNCIL OF INDUSTRIAL DESIGN

CAMERA PRESS

Choosing the fitments

When considering any particular room, bear in mind that several types of distribution and fittings may be better than just one. There are some important general points to bear in mind; glare should be avoided, and lamps in the fittings should be shielded from view. Too great a contrast between the light fitting or lighted object and its surroundings is likely to cause discomfort, and sometimes eyestrain and headaches. Even the flame of a candle on a dinner table can cause considerable glare if there is no background lighting, because of the extreme contrast between the candle flame and the surrounding darkness. Television should not be watched in a totally dark room for the same reason.

When choosing your lighting fitment, do be careful to consider the effect of the lighting and not just the look of the lamp. A beautiful standard lamp that will grace the corner of your living room may have a totally impractical effect, casting huge shadows, or giving a glaring light. So often additional lighting fitments are just added to a room, apparently haphazardly, around the furniture. You end up with an uninteresting effect and yards of trailing leads.

Daylight is constantly changing in colour and intensity. Similarly, artificial lighting should provide contrast. Pools of light, reflected light, lower levels of lighting—all can add interest to a room and break up different living areas.

Lamp shades are also important. Large ones allow greater diffusion of light. Their colour will greatly affect the light, too. 'Warm' colours—reds and yellow tones—are generally considered attractive. Blues and greens can be used to great effect for special positions provided they are backed with warmer shades in the rest of the room.

Table lamps should always have shades large enough to cast a pool of light over sewing,

THE ELECTRICITY COUNCIL

***Left.** Softly diffused light enhances a distinctive modern setting by emphasizing the contrasts within the room. It is reflected off the walls to fall downwards.*

book or table.

Fluorescent fittings are particularly suitable for providing a good economic source of light. But be careful about choosing the colour of fluorescent tubes as they can easily give the wrong effect. 'De luxe warm white' tubes are the most suitable for domestic purposes because they are the closest tube colour to tungsten lighting, and also reflect colour well. Nearly all fluorescent tube manufacturers produce tubes specially designed for giving good colour, but some can appear much colder than 'de luxe warm white'. If you want strip lighting in the living room or bedroom, a particularly warm tone of light—the red tones—may be needed. Tinted fluorescent tubes, such as pink or peach give a soft, attractive light. The 'colder' tones, however, can be useful to bring up blue or green shades of decor or foliage.

Tungsten tubes have none of the disadvantages associated with fluorescent lighting, such as buzzing, flickering and delayed starting, but they are more expensive both to buy and run.

Although principally intended for the shop fitting and display market, spotlights can provide interest and atmosphere in a room. (But be careful about positioning them, or someone may get the impression he is being put through the third degree!) The development of tungsten halogen lamps (probably more familiar to most people in projectors), has enabled tiny spotlights to produce a very high output.

Tracking light systems, which can hold several spotlights at different angles, also have adaptors for taking low load appliances such as food mixers, sewing machines or portable electric drills. Because of the possibility of overloading too many fittings into the track, it is preferable to have it installed by a qualified electrician. Should you decide to install one yourself, check that the current carrying capacity of the cable is adequate for the purpose, and that an appropriate fuse is provided in the circuit.

Another by-product of the shopfitting market is the general availability of colour spotlights. You have only to think how colour television can portray the whole colour spectrum simply by the projection of red, green and blue onto the screen to realise the enormous potential of colour lighting. With imagination—and discretion—you can change the whole atmosphere of a room by the use of colour filtered light.

Wall brackets and wall mounted fittings are particularly useful for such rooms as the living room, where you need semi-indirect light from above and behind.

Using daylight

Do not forget daylight. It is as much a part of planned lighting as the most sophisticated artificial light. Where the opportunity arises of influencing or dictating the positioning of windows, these should be carefully planned to achieve a good balance of light in the building. Very large windows are not necessarily the answer. Large rooms are for the most part better lit with more than one window, preferably in different walls. But the positioning of windows involves far more than simply its effect on the daylighting of the interior; orientation, view and the external design appearance of the house are also major considerations.

As well as the actual positioning of windows, think of the window space as potential decoration through the curtaining and colour of the window frame. A deep room with a single window, be it large or small, should not be decorated in dark finishes if the room is habitually used in daylight. Remember, too, that glare from an over-large window can be softened by hanging fine net curtains in front of it.

Do not neglect the exterior of your house. A pretty garden can look dramatic and glamorous when carefully lit and give a visual extension to your living room. Lights by the side of the drive reveal it quite clearly at night.

Lighting can be fun, formal, and flattering. Experimenting with it is fun, too, but the secret often lies in simplicity. A single spotlight focussed on a favourite picture is a simple idea, but can be more effective than a dozen badly placed standard lamps. Planning lighting carefully can give your home a new dimension.

Planned lighting-2

MICHAEL BOYS

Because of the technicalities involved, many people tend to fight shy of making drastic changes to the lighting in their homes, preferring to leave it to the experts. But traditional lighting in the home is usually conservative, and often inadequate. By exploiting the potential of the many different fitments that are readily available, you will be able to achieve both effect and efficiency in your lighting system—and without completely rewiring your home.

Unless you are lucky enough to be in on the planning stage of building a new house, you will probably find that your home has been wired unimaginatively. Builders tend to be conventional and provide only the standard lighting points in the traditional positions—a centre ceiling light, one wall plug per room, and perhaps a couple of wall light outlets on either side of a fireplace or in an alcove. It is all too easy to allow your choice of lighting to be moulded by what is there—and obvious—rather than plan out exactly what each individual room needs. After all, planning takes time and energy. But if you can find enough of these to bother, the rewards will be considerable.

When choosing your lighting fitments, try to decide what you want before you go shopping. Department stores tend to be discouraging to new ideas. Their lighting departments are usually a blaze of fittings, all switched on at once, with little attempt to demonstrate individual effects. It is hard to visualize what a specific fitment will look like away from all the others in the shop and hanging in your home. Confused and unsure, one can so often end up with buying the same style fitment or lampshade as before, with the feeling that it is better to be safe than sorry.

Spotlights

Although a new lighting scheme has ultimately to be planned for the whole house, it can be tackled gradually. One of the easiest, cheapest and most effective experiments you can try is to purchase and place a spotlight. Theatrical designers and shop window dressers have realized—and exploited—the dramatic potential of spots for many years. A figure on the stage, or a garment in a shop window, can be made arresting and dominating by the simple expedient of focussing a spotlight onto it.

Spotlights are now supplied specifically for the home and, used with a little imagination, they can do much to enliven a room. Because they are fully adjustable, both vertically and horizontally, even one lamp, carefully placed, can be used to illuminate a favourite painting, for example, and then turned to throw intense local light on a work surface such as a desk. Apart from wall fittings, they can be placed on the floor as elegant uplighters, fitted to standing frames or ceiling tracks, or clamped to a bedside fitment for a reading lamp.

Domestic spotlights are mostly small, neat and unobtrusive. The light and not the lamp catches the eye—or should if it is properly placed. As well as ensuring that they do not give a blinding light, spots must also be positioned away from fabrics or any inflammable material as they produce a considerable heat.

Normally, spotlights are fitted with special, internally silvered reflector lamps, but you can also buy general service lamps adapted with clip-on auxiliary reflectors. Versatility is the great merit of spotlights. Installation is simple: they screw into place, and are wired like any other wall or pendant light. Some have push-button switches, so they can be wired into an unswitched circuit, without the need to run new cables back into the switch for other lights in the room. If you want to fit them to the ceiling, spotlights should be sited one to two feet from the wall. Placing them depends on what you want to illuminate. Spotlights should 'spot' something—be planned to pick out one area or object. Installed at random, much of their effect is lost.

Fluorescent lights

Fluorescent tubes are often considered boring, dazzling and useful only for kitchens. This need not necessarily be the case. They can have many effective and efficient uses in the home and are always worth considering because of their cheapness to run.

The slightly chilling effect they sometimes have can be eliminated by using one of the warmer shades such as 'De luxe warm white'. 'Warm white' provides a high light output (measured in lumens per square foot) and is very efficient in large kitchens. It is 'sharp' enough for work surfaces where efficiency is important, and yet warm enough to be comfortable to work in. 'Natural' is closest to daylight but is probably the least acceptable for domestic use since it gives a cool, blue-white appearance.

Daylight itself reflects 'true' colours because

THE ELECTRICITY COUNCIL

it contains all the colours in the spectrum: red, orange, yellow, green, blue, indigo and violet. Artificial light, in general, contains a much narrower range of colour, which is why fabric shades, for example, look so different under department store lights. Artificial light gives either a cold or a warm effect and you must know what is best for your needs. Candlelight or firelight contains more natural red/orange and gives people a healthy, flattering glow. Fluorescent tubes tend to have bluey tones which give a colder effect. This is why they are often recommended for use in purely service rooms where you need a good light rather than an intimate atmosphere.

Despite its occasional chilliness, fluorescent light has much to recommend it. Because the length of tube eliminates light shadows, a 40 watt tube gives approximately the same amount of light as a 150 watt metallic filament bulb and spreads the light over a wider area. And with an estimated life of 7,500 hours, a tube lasts at least seven times longer than a bulb.

Another advantage is that the tube never gets overheated; you can even fix one behind a pelmet with no danger of your curtains catching fire. Fluorescent tubes are available in a range of lengths from 1ft to 8ft and your choice of length depends on the output needed. For example, a 15 watt, 15in. tube would be ideal for lighting a bathroom mirror, and a 40 watt, 4ft tube would be fine for a small kitchen.

The installation of fluorescent fittings presents no particular difficulties. The body of the unit is screwed to the wall or ceiling through distance pieces, designed to provide a space of about ¼in. between the back of the unit and the surface to which it is fitted. This is to allow some circulation of air round the tube. The

Left. *Dramatic use of whole-house and surrounds lighting, with spotlights under the roof and special porch fitments to give a welcoming blaze of light.*
Below. *Living area enhanced by the use of several types of lighting. The central light has been extended to hang at any chosen spot, plus extra spotlights for reading.*

wiring is equally straightforward, being clearly marked on the terminal block fitted inside the unit. Some fittings have push-button switches so that they need not have a switched electrical supply and these are particularly useful for fitting over a bedhead.

While unshielded tubes are not ugly, they have a stark, utilitarian air, and you may prefer to use them only in less sophisticated areas, such as a workshop or kitchen. The tubes can be made more attractive, however, by fitting them with diffusers, although some output efficiency may be lost as a result. There is a wide range to choose from: extruded reeded plastic, natural raffia, and various fabric finished diffusers. They are designed to give a soft upward light, with a stronger downward light to illuminate areas of special interest. Diffusers usually clip to the tube itself, or to the fitting, and are easily removed for cleaning.

Ceiling lights

Fittings that are mounted on the surface of the ceiling, or even recessed into it are a comparatively modern development. They can be spotlights, or more conventional fittings with diffusing glass or louvres which are excellent for general lighting. Spots are ideal for reading because they provide beamed, concentrated light; the light shines directly onto the book and the reader is not sitting in his own shadow. These lights can be grouped to provide high light intensity where it is needed, giving reflected light throughout the rest of the room.

Not many people will want to start cutting holes in their ceilings to accommodate recessed light fittings. A less messy and inconvenient alternative would be the installation of a false ceiling beneath the existing one. This may be worth considering if your present ceiling is either badly cracked or very high (more than 8ft 6in). Tackling a whole ceiling can be an ambitious project, but you may be able to fit a false ceiling across part of a room. This is particularly easy in a house that has a reinforcing beam in some of its ground floor ceilings, supporting a bay or ground floor extension. This would enable you to fit a false ceiling to the

MICHAEL BOYS

depth of the beam. With two or three concealed lights built into your new ceiling, there would be no ugly cables showing and there would be the effect of softly diffused light.

Try to forget about the traditional habit of placing a light in the centre of the ceiling. Several lights dotted all over the room are more interesting than a single fitting, although an off centre ceiling light can only be really efficient if other lights are available to illuminate the rest of the room. Greater flexibility can also be achieved with ceiling lights if they are fitted with a dimmer switch. These simply replace your present switch and enable you to raise or lower the level of lighting by simply turning a dial. In this way, you can, for example, tone down powerful reading lights for relaxing.

New ideas

Even in the more conventional lighting fields —such as ceiling mounted or suspended fittings—there is a wide range of choice and something to suit every taste.

The bayonet-type fitting, suspended on a flex, remains the most popular. Designs are available to suit the mood of every room and they have the advantage of being extremely easy to fit. The flex need no longer be the ugly, twisted, silk-finished type. A single tubular plastic insulation looks more attractive and can be wiped clean.

Rise and fall units—ceiling suspended lights which are height adjustable—are also useful and versatile. When used in the 'up' position they provide general lighting. In the 'down' position they throw a pool of light into a limited area, creating an intimate atmosphere. They are most commonly used for dining table fittings and are particularly suitable for this, although they could also be used over a desk top or sewing machine, for example.

Many people choose a standard lamp for a reading light or to cut down glare when watching television. These will be more efficient if fitted with a large shade, which gives a wide pool of light. Small shades also tend to make

COUNCIL OF INDUSTRIAL DESIGN

Opposite page. *This unusual table light in the modern style, combines adequate light with an intimate atmosphere by covering the surface with soft light.* ***Left.*** *This modern standard light is contoured to merge with the comfortable modern look of the room. Its adjustable head provides for variations in mood. Whilst the spotlight can highlight the pictures whilst leaving the room in shadow.* ***Below.*** *As you can see unshielded ceiling lights need not necessarily be ugly. These hall lights give a soft, welcoming light to the hall without uncomfortable glare. At the same time it is essential to provide adequate lighting on stairways and in passages, for safety reasons.* ***Right.*** *A wall-mounted desk lamp which can be adjusted to several angles, throwing intense local light on the work area.*

ELIZABETH WHITING

HEIDEDE CARSTENSEN/GRUCO-MOEBEL

HEIDEDE CARSTENSEN/GRUCO-MOEBEL

bulbs overheat, thus shortening their lives, because they prevent the free circulation of air around the bulb.

How much light do you need?

The various activities carried out in the average home require not only different types of lighting, but also different amounts of light. In technical terms the requirements are expressed in *lumens* per square metre. (To convert them to lumens per square foot, divide by 10.)

For general use a living room should have a light intensity of 100 lumens per square metre (10 per square foot). Light reading requires 200, prolonged reading (or studying) 400. Sewing or close work calls for 600 lumens per square metre. You could achieve this necessary versatility by having a ceiling light fitting which provides sufficient light for general purposes, supplemented by spots or table lamps. Additional light will be needed near or above chairs or work surfaces.

The average requirements for other rooms are Kitchens—200 lumens per square metre (particularly over work surfaces); bedrooms—50 lumens per square metre with provision for a 200 lumens headboard light for the late night reader. Most other rooms, including landings and halls need about 100 lumens. Stairs, at their treads, ought to have 50-100 lumens to be absolutely safe—with control switches at both the top and bottom of the staircase.

It is simple to calculate the wattage required to provide these standards of lighting as long as you know the efficiency with which the lamp converts electrical energy into light—in other words, the number of lumens per watt. A fluorescent tube provides about 50 lumens per watt. So, for example, if the light intensity needed in a kitchen is 20 lumens per square foot the wattage required using a fluorescent tube will be $20 \div 50 = 0.4$ watts per square foot. Thus, one 40 watt fluorescent lamp will provide the equivalent of 20 lumens per square foot over an area of 100 square feet (0.4 x 100). Metallic filament bulbs are far less efficient than fluorescent lamps, producing approximately 12 lumens per watt.

Daylight has hundreds of lumens per square foot but it is generally easy on the eye because it is so evenly distributed.

Rewiring

Unless you are building a new house or having your home completely rewired—when you can merely give the electrician your instructions—you may find that efficient, effective lighting involves some rewiring. However, do not be too ambitious. It takes an expert to rewire a house, especially the complicated rewiring you may need to achieve your best planned lighting. Use your own ideas but, unless you have every right to consider yourself an expert, leave the actual rewiring to a professional. Many useful improvements can be made very simply by adapting your existing circuits—preferably with the minimum of damage to your decor.

For example, a lounge which is fitted with one central pendant light and a single power point has very severe lighting limitations. But a few simple changes can work wonders. Rid yourself of the centre light and replace it with three ceiling mounted spots in the darker end of the room, away from the window. These could be served by a dimmer switch, and supplemented by a table lamp, and/or a fluorescent tube concealed behind a pelmet or in a shelf recess. Or try splitting one lighting outlet, such as the wiring for a wall light, to provide wiring for two mounted spotlights. These could be served by the same switch. If you decide to mount the cables on the surface, look for unobtrusive routes. (Burying them behind skirting boards and plaster tends to upset your decor.) You could run them behind curtains, along the top of a skirting board, or down the corners of the room. Securing clips with hardened steel pins—such as Dylon clips—are small, easy to use and make a neat job of fixing cables to walls.

When you plan the lighting of any room, aim for variety of effect. Daylight is always interesting because it is constantly changing in intensity. Artificial light should try to be just as interesting, just as adaptable. With a little care, your lighting will add greatly to the efficiency, mood and atmosphere of your home.

THE ELECTRICITY COUNCIL

***Left.** When planning lighting for halls it is important to bear in mind the need for adequate light for safety and for vanity purposes. This lighting arrangement allows light to shine on the face whilst cutting down on glare.*

GRAZIA NERI

Above. *Apart from bringing out the colour and texture of the timber most effectively, the concealed fluorescent strip gives adequate light to the work surface in this well planned kitchen.*

Lighting for kitchen and service areas

Never underestimate the importance of good lighting in the home; too many people put up with a far worse standard of lighting at home than they would tolerate at work. However well decorated and designed your house may be, without well designed lighting it will never be really comfortable. A well thought out lighting system is especially important in the kitchen/service area, where good light is essential for food preparation and cooking. Bad lighting can lead to irritability and mistakes—very dangerous in the kitchen.

When you move into a new house you are likely to take existing lighting provisions for granted, and when it comes to designing the layout of your new home, accept the position of the old lighting points without question. Also when you are planning extensions or conversions in your home you're more than likely to stick to traditional layout for lighting—a central ceiling fitting with maybe one or two wall lights. The electrical outlets in modern houses are usually kept to a minimum to keep down building costs. This tends to produce a standard of domestic lighting which is well below recommended levels.

People often accept a worse standard of

BILL McLAUGHLIN

lighting at home than they would do in their workplace. It could be argued that lighting is not so critically important at home as in a factory or an office. This is not true of the kitchen. You do as many intricate and demanding jobs here as you would for any employer.

The importance of good lighting

Good lighting in the kitchen is essential—it makes housework safer, easier, cheaper and more efficient, as well as improving the look of your kitchen.

Safety will be much improved by adequate well placed lighting that will avoid you working in your own shadow, or that of a cupboard. This will reduce the possibility of accidents. Switches should also be sensibly positioned. There should be a main light switch close to every entrance. The rocker switch is easier to operate than the conventional up and down type—especially when your hands are full. Also, you'll find it easier if all switches are placed at door handle level, although this is not a good idea if there are young children around. Switches in areas where your hands are likely to be wet, for example by the sink or in the laundry, should be operated by a pull-cord from the ceiling. All switches should be easily accessible without reaching, stretching or moving things out of the way.

Above. *For a dining area in the kitchen a pendant, such as the one shown, is ideal. Note that although the light is very bright it is well diffused—giving comfortable illumination.*

Eye strain and general fatigue will increase the possibility of accidents. This will occur if the general light level is too low. It can also happen if lighting levels change too dramatically from area to area of a room. Your eyes are capable of adjusting to extreme changes in lighting levels—for instance bright sunlight can be anything up to 30,000 times brighter than moonlight—but this adjustment takes time. If you move quickly from job to job—each in different light—eye strain will be worsened. Vision may even be temporarily affected.

Economy and efficiency

A well designed lighting scheme should incorporate a number of different lamps, each on a separate switch. This will be a little more expensive to install than a conventional system but you'll save money on your electricity bills in the long run by being able to switch off unwanted lights. There is an added advantage in a kitchen/diner because you can reduce the lighting level while you eat. An alternative to switching lamps on and off is to fit a dimmer switch. Dimmers can be easily fitted by the home electrician. Dimmers enable you to reduce or increase the light level of a single lamp by operating a control at the switch.

Well positioned and easily controlled lighting of the correct intensity is bound to add to the efficiency of a kitchen. You will be able to work quicker and to a higher standard. Good lighting will also enhance the rest of your design work by emphasising colour schemes, cupboard units and kitchen furniture. It will be an aid to cleanliness, and make the kitchen look more spacious. All this increases efficiency by adding to the enjoyment of work.

Working out lighting requirements

It is possible to arrive at some approximation of the amount of light your particular kitchen needs. This is desirable if you are building an extension or modernising your kitchen. It's also a good idea to check the light levels of your existing kitchen, where there may be something lacking which you can't quite define.

In fairness to yourself, it's worth checking even if your lighting seems adequate. It is usually a relatively simple matter to check the lighting in your kitchen because of the fixed nature of most kitchen furniture, and the uniformity of jobs done in the kitchen. This

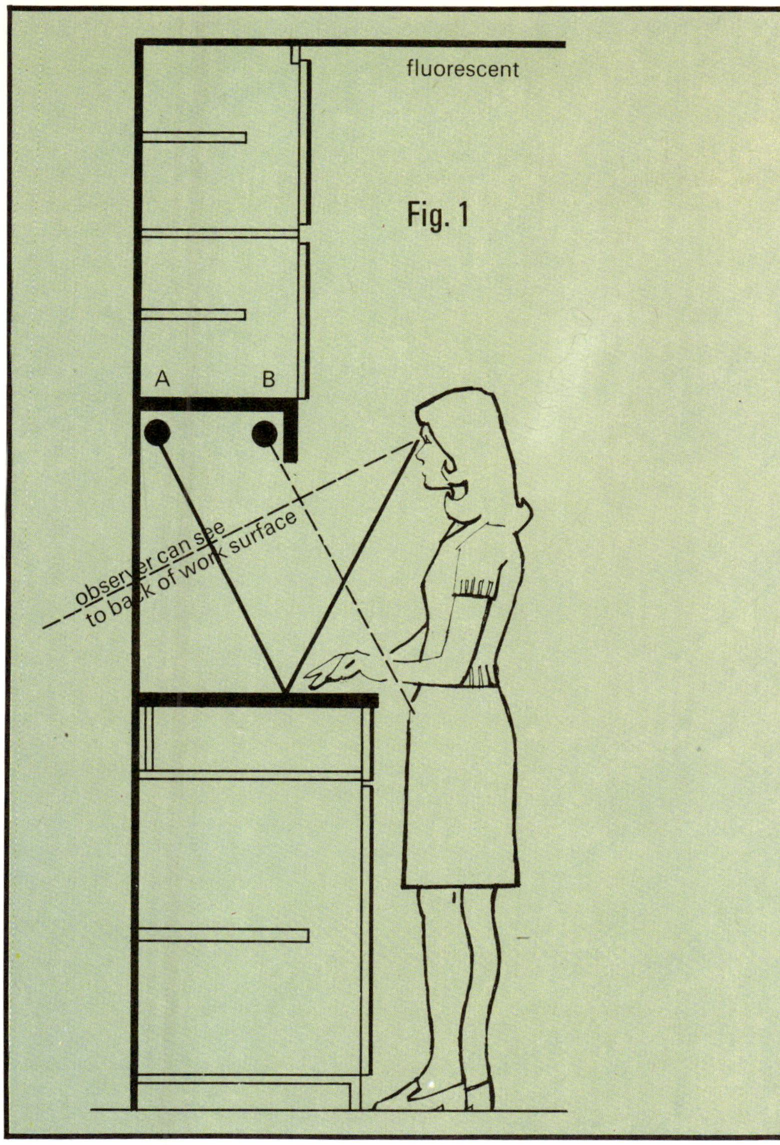

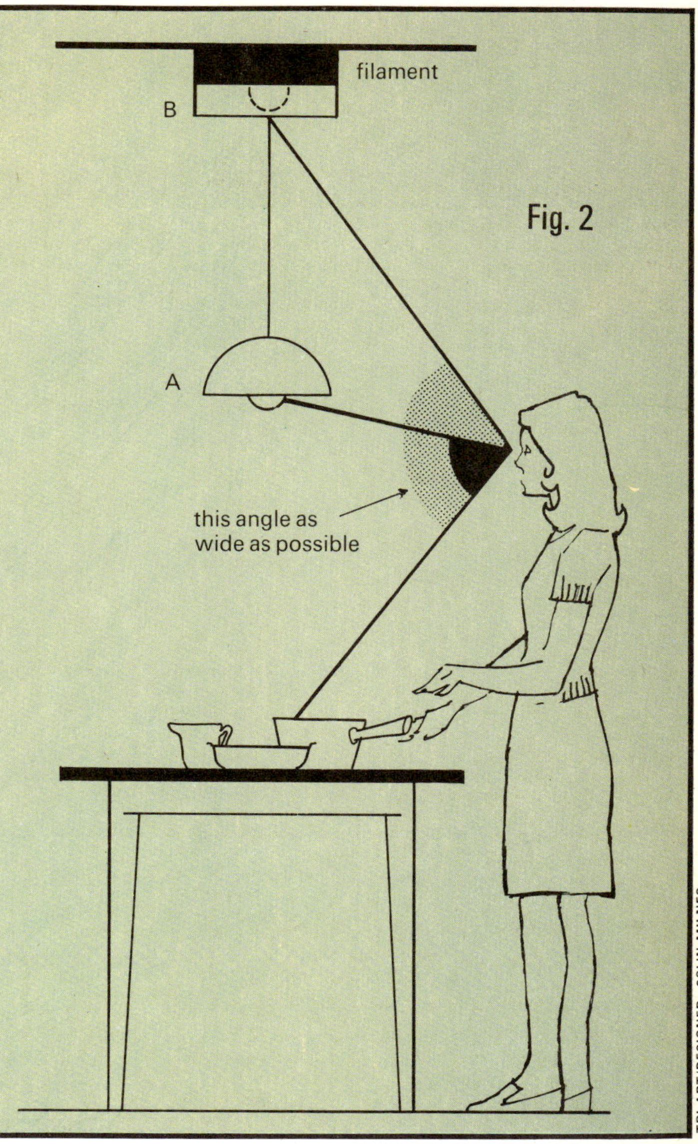

TRI-ART/DESIGNER: COLIN MILNES

***Fig.1* (above). *A.* Fluorescent tubes in this position will throw too much light on to the back wall—the reflection will cause dazzle. *B.* If the fluorescent tube is fitted here, the result will be to give even light to the work surface without causing dazzle.**

***Fig.2* (above). *A.* A pendant light in this position will be far too close to the task area for comfortable working. *B.* The light source should be as far away from the task area as possible. It may also be necessary to fit a diffuser to avoid glare.**

means that the lighting needn't be as flexible as in other parts of your home. First of all it will be helpful to explain briefly how light is measured and to point out the main differences between fluorescent and filament lamps.

Light is measured in lumens per square foot. A 'lumen' is the amount of light one candle gives one foot away from the flame. An average daylight reading, away from direct sunlight, is somewhere between 500 and 1,000 lumens per square foot. The recommended light level in the kitchen is 20 lumens per square foot.

Fluorescent lamps are normally available in the form of tubes of varying lengths and output. Fluorescent lamps require a starting mechanism which is usually built into the fitting. The actual starting process of a fluorescent is less economical than a filament lamp, but once running it has approximately three times the lighting efficiency of the filament, and consequently is cheaper to run.

Filament lamps for domestic use normally come in the form of the familiar pear shaped bulb. Compared to fluorescents they give off a yellowish light and burn at a much higher temperature. They give a more direct light than fluorescents and therefore cast stronger shadows. For this reason they are less effective for concealed lighting techniques.

Positioning of fittings

Once you have decided how much light you need in your kitchen, the next step is to plan the number, type and positioning of the fittings. As with all designing, your decisions will be influenced by various practical limitations as well as by your personal preferences. The main points to consider are daylight factors, general work areas, the shape of the room, and body sizes in relation to work areas.

Daylight factors

Most of the time you spend in the kitchen throughout the year is in daylight hours. Before planning your kitchen lighting requirements take note of the size and position of the windows, then wherever possible provide artificial light from the same direction to give consistency. East facing windows are good because they receive the morning sun. South and west facing windows receive the sun later in the day when it is usually too strong for comforable work. Venetian blinds offer excellent control but their horizontal slats gather dirt quickly—involving you in a lot of cleaning work. Vertical slats are available, although they are not very common. In some cases external shades may offer the best solution. They shelter the room from hot direct sun but allow low winter sunshine to enter the room.

Where there is only one window in the kitchen it should not face north as the room will not receive any direct sunlight at all. The closer a window is to the corner of the room the more light will be cast on the adjacent wall. Windows between wall cupboards and the ceiling will allow reflected daylight to enter the kitchen. A good direct light source in which to work will be provided by a window positioned between wall cupboards and work surfaces.

Daylight gives more than enough light for the kitchen. Usually, however, only a small proportion of this light actually enters the room. The

PHOTO: JOHN VAUGHAN/ENGLISH ROSE/WESTINGHOUSE KITCHEN

area immediately inside the window will have a light level of only about a tenth of outside light intensity, and the light level at the other side of the room will again be only about a tenth of the level immediately inside the window. The area farthest away from the window should have additional local lighting for use during the day. This will be even more necessary if you work with your back to the window.

Other influences

If the light in your kitchen is of equal intensity throughout it will result in the room having a flat, uninteresting appearance. You should aim at two main light levels in the kitchen—general and specific. General light is the overall light provided by the main source, while specific light is extra light provided for particular work areas. Proper positioning for specific light sources can be determined by making a list of all the activities you are likely to carry out in the kitchen. Allocate areas for these and plan your highest level of illumination accordingly. General lighting should be at least 15 lumens per square foot, while specific lighting should be about 30 lumens per square foot. This gives an overall average of 20.

Be very careful to avoid glare around work areas. This will occur if the background is brighter than the job in hand or if it is lit by a naked light. Lights fixed under wall cupboards designed to illuminate work surfaces should be concealed behind baffles to prevent glare. The baffle must be fixed high enough that the back of the work area is not obstructed. The possibility of reflected glare remains, but this will be greatly reduced by careful positioning of the lamp.

Glare will also result if the light source is placed too close to the job. The angle formed at

Above. *Clever use of light accentuates the uncluttered design of this kitchen. Adequate overall light is combined with sensibly planned illumination for the work surfaces.*

eye level between the light source and the particular task should be as wide as possible. You can reduce glare further if necessary by using a fitting with a diffuser attached.

If you require additional lighting, install lamps in cupboards and alcoves. The layout of your kitchen/service area will have a bearing on this. Cupboard lights are best operated by an automatic door switch as used in refrigerators. This eliminates the chance of leaving a light on behind a closed door.

For a successful scheme you should design the lighting along with the interior structure and finish of your particular kitchen/service area. What is an outstanding success in your friend's house will not necessarily work out in your own home. Ceilings and cupboards, along with recesses, should be designed to incorporate concealed lamps. Colours and textures should be chosen to suit your lighting scheme.

It is important to familiarize yourself with basic lighting techniques and the range of equipment available. This knowledge will help you to make sensible decisions based on what is and what is not possible in *your* kitchen. Once you are familiar with standard wattages in relation to lamp sizes and special fittings you can apportion your total output for the kitchen according to general needs and work areas.

You should mark down all your lighting requirements on a drawn plan of the kitchen, before beginning any practical alterations. Your drawn plan should indicate the height of fittings from the floor where appropriate. Work out how best to incorporate your lighting scheme into the rest of your wiring requirements.

When choosing fittings you should concentrate on the quality of light provided rather than the appearance of a particular fitting. The number of designs available should enable you to find a good quality fitting which satisfies your tastes at the same time.

When designing recesses to house concealed units, remember that the fittings must be accessible for cleaning and replacement. They must also be adequately ventilated to prevent overheating which is both dangerous and wasteful.

When decorating your kitchen/service area choose pale colours over large areas for maximum light efficiency. Avoid having very shiny surfaces close to concealed lighting as reflected lamps will spoil the effect as well as increasing the chance of glare.

Concealed lighting

There are several concealed lighting techniques which can be employed in the kitchen without reducing efficiency.

Old kitchens with high irregular ceilings can be given a clean modern appearance with the addition of a suspended illuminated ceiling – page 44 shows you how. There are a number of standard types available. Most of them are made up of a number of adjustable suspension rods to which a network of inverted 'T' sections are fixed between 12" and 18" (300mm and 450mm) below the existing ceiling. A wide variety of diffusing panels are available. These simply rest in place for easy removal. The fluorescent lamps are fixed to the existing ceiling.

Cove lighting is ideal for kitchens with wall cupboards that do not reach the ceiling. A baffle is added to the tops of the cupboards and fluorescent tubes are fixed behind. The light is reflected from the ceiling. Although this technique will not alone provide adequate working light it will appear to heighten a low ceiling and add a sense of spaciousness to an otherwise cramped room.

Recessed filament light fittings are unobtrusive and are available in many different styles. This technique is particularly suitable for service areas with low ceilings, or as a complement to rooms where the main light source comes from fluorescent tubes.

Lighting a dining area

If you have a dining area in your kitchen your provision for lighting will be very much a matter of personal preference. However, assuming that you are likely to use this area for a number of activities, you should provide a feature pendant over the dining table with a maximum output of about 300 watts. This light should be operated by a dimmer switch for increased flexibility. In addition to the main pendant you could provide wall lights, operated by pull cord switches. The number of wall lights you fit will depend on the size and the structural features of your dining area. You could also consider fitting fluorescent lamps for a shelf displays. This will look most effective when used on open wall shelves or a Welsh dresser. Sensibly placed shelf lighting will reduce the often disturbing contrast between the kitchen and dining area lighting.

TRANSWORLD/SCOOP

Lighting for living rooms

The living room is the most important room in your house—and the way you light it makes all the difference to how it looks and the use you make of it.

Because you take specially great care when it comes to planning the layout and decor of your living room, you will want lighting that complements the decorative scheme and reveals it as effectively—although very differently from—daylight. You use this room for a multitude of activities from watching television to sewing, reading or even sleeping and very often more than one of these is going on at the same time. Because of this you should try as far as possible to design the layout of the lighting in your living room to cater for your family's individual needs.

Imaginative and sensible use of light is central to the success of your living room both from the decorative and functional points of view. Your best thought out decorative scheme will not succeed if, for example, your wife cannot read a book comfortably in the subdued light you require to watch television. It's always best, therefore, before you start decorating to plan your exact lighting requirements. If you are in the lucky position of starting from scratch, while your future home is in the process of being built, be generous with your provision for power points. Even if you think you may have no use for them at the moment they will save you extra trouble and expense when installing additional light fittings.

When you have decided how to arrange the lighting in your living room the next task is choosing from the vast range of different light fittings on the market. This need not be difficult if you stick to two basic guidelines: first, decide firmly what purpose any individual fitting is supposed to fulfil, and second, choose a fitting that both meets your requirements as exactly as possible, and is attractive to you. It's very easy when choosing light fittings to fall in love with one that may look wonderful on display in the shop. Beware of making a snap purchase as the fitting, however beautiful, may not be intended to do the job you want. Make sure that the function of the light fitting, as well as its looks, match your living room.

You will derive the greatest benefit from your lighting system if you try to achieve a compromise between the general need for overall illumination and individual needs for extra light in certain areas of your room. The system most comfortable to live with is one where lights both for general and individual needs are provided and successfully used. If, for example, you place a standard lamp in one corner of your room near an armchair or studio couch and keep the main ceiling light at a slightly subdued level you will find the light more comfortable for general illumination than just the main ceiling light on full power. At the same time the standard lamp gives you a bright 'pool' of light for reading or sewing. Similarly if you fit wall lights in alcoves they can provide bright corners in a generally restfully lit room.

Safety is of first importance when designing your lighting system. Make sure that all your power points and light fittings are properly earthed. With new houses and new fittings this is now a legal requirement. All light fittings must be either earthed or have double insulated flexes. If you have old fittings check that they meet these conditions and replace ones that don't. Never fit wall lamps to a wall if you have the slightest suspicion of the wall being damp.

Main ceiling lights

The most important light in your living room is likely to be the main ceiling light, although in some houses there is no main ceiling light and its job is done by fitted wall lights. The main light is intended to give a general illumination to the room when you require an overall light to carry on with a number of activities. Forget, however, about the widely held 'brighter the better' idea. It is far better to use a number of fittings to provide illumination for particular tasks than to rely on one really bright central light.

The output of the main light should be calculated against your colour scheme, the number and output of other lights and the size of your living room. If the decor of your living room is all white a very bright main light will soon become oppressive and will succeed only in giving you a splitting headache. The same will apply if your living room is very small. People's sensitivity to light varies enormously so it is difficult to lay down hard and fast rules for the output of main lights. However, if you have a small living room a total output of 150—200 watts should be enough for the main light. If your living room is on the larger side proportionately more output will be required; 250—300 watts, possibly split between two or three bulbs, will provide enough general light for a sizeable living room. In a large living room, two or more ceiling lights are often better than a single fitting.

There are a great variety of main light fittings for you to choose from. The most common is one or more bulbs suspended by a flex from

Right. *Small, wall fitted spotlights make a beautiful corner fitting of these otherwise ordinary bookshelves.*

RICHARD EINZIG

the ceiling with the shade designed to throw light over a wide area of the floor. This kind of fitting is standard in most homes and can be made very attractive by the choice of a suitable shade. You must bear in mind one point when choosing a shade for this kind of fitting and that is that the shade should diffuse the glare but not reduce light. A number of shades, while often being quite attractive in themselves, do not satisfy this condition. Of the ones that do one of the cheapest and best looking is of the globe design. This shade is collapsible and made of white paper. It is simple to fit but be careful when handling it so as not to tear the paper.

An automatic ceiling housing which reels the flex inside itself will make it possible to raise and lower this central light and make for more flexible use. An inexpensive way of arranging this is to replace the existing flex with a much longer piece. Spools can be obtained to wind the flex around. All you need to do then is to put the spool inside an adapted ceiling housing and you can alter the height of the light at will. Lowering the height of the light will have the effect of darkening the upper parts of your living room and concentrating light on the floor. This achieves a subdued effect.

You can attach the same kind of fitting directly to the ceiling and this can be a good solution for rooms with low ceilings—a common feature of newer houses. In other cases you may wish to fit a shade that throws light up onto the ceiling to give a softer illumination. Lighting a ceiling or a wall in this way is a simple method of providing general, glare-free illumination.

If you have, or intend to fix, a false ceiling in your living room main ceiling lights should be recessed into the new ceiling. Attractive recessed light fittings are available incorporating ordinary tungsten bulbs. Most types of recessed fittings are circular in shape but you can obtain a wide variety of other styles including square, rectangular, oval and even hexagonal. You also have a certain choice in the depth of the recess, although in the case of main ceiling lights the recess should be kept fairly shallow.

One method you can use to provide a main ceiling light which is especially suitable for very large living rooms is by the use of enclosed strip lights. These are available in various sizes and consist of a number of strip lights placed parallel to each other and enclosed within a glass or plastic white opaque case. You will find the most common shape to be similar to that of a small paving stone, though other shapes exist. If fitted into a false ceiling this type of light looks both neat and impressive in the larger room. The average size of the fittings and the brightness of the strip lights makes this fitting unsuitable for use in the smaller living room.

Table lamps

Table lamps are available in a wide variety of styles ranging from traditional designs through to the modern-looking stainless steel and enamelled models which were first popularized in Denmark and other Scandinavian countries as long ago as the early 1940s.

Think of table lamps as a means of providing 'pools' of light in appropriate areas of your living room. Sometimes they will add to the main light, at other times you will want to use them by themselves. As mentioned earlier you should attempt to plan your lighting requirements to cater for the individual needs of your family. Don't litter your room with table lamps in every available odd corner but try to position them where they will be needed—on a low table near an armchair or couch, for example. A well designed table lamp will provide adequate light over a limited area for reading or sewing while making no significant difference to the subdued light you may require for the rest of your living room.

If your living room is on the large side you may decide to apportion certain areas of the room to different uses. This is best done by means of room dividers. What types of activity you decide should be carried out in different areas of the room will determine your lighting requirements. Placing a table lamp on a high shelf of a room divider will give you valuable extra light in areas of the room where the dividers, by casting a lot of shadow, cut down the effectiveness of your main light source.

Standard lamps

As with all types of light fitting you will find that a wide variety of standard lamps are available. It is a mistake to choose a standard lamp for purely decorative reasons, although you will obviously be influenced by the beauty of any particular lamp. After the main ceiling light a standard lamp will provide you with the most general illumination and in many respects a standard lamp will give a more satisfactory light for everyday use than the main ceiling light. Properly positioned a standard lamp, fitted with a suitable shade, will provide a reader with excellent illumination (see diagram **B**). Also you will find that it will light your living room quite adequately, and at the same time will not be glaring enough to give you eye strain or headaches.

Ideally, to give the maximum light, a standard lamp should be positioned so that there is a wide area of floor all around. Failing this you should place it near a wall painted in white or another light colour. Not only will this look very attractive but the light reflected from the wall will give more illumination to the room.

If you wish to achieve overall control of the light for your living room—moving simply from bright lighting to subdued—the easiest and most inexpensive way of doing it is to fit a dimmer switch. This replaces the switch operating the main light and enables you to alter the brightness of the main light. You can also use it to control more than one light. Two designs of dimmers are commonly available. One consists of a panel of switches. Turning these switches on one after another results in an increase in brightness by stages. When all the switches are turned off complete darkness results. As the switches are turned on light intensity increases until full brightness is achieved.

The second design is the more versatile of the two and offers you infinite variety in light intensity from full brightness to complete darkness. It replaces the switch with a knob—turning which, by turning in a clockwise direction, switches on the light and then increases its brightness. You must make certain that you

Bottom left. *Three dimensional murals and tapestries are brought to life by a spotlight fitted as shown in diagram* ***A****. Diagram* ***B*** *indicates the ideal position for a reading light.*

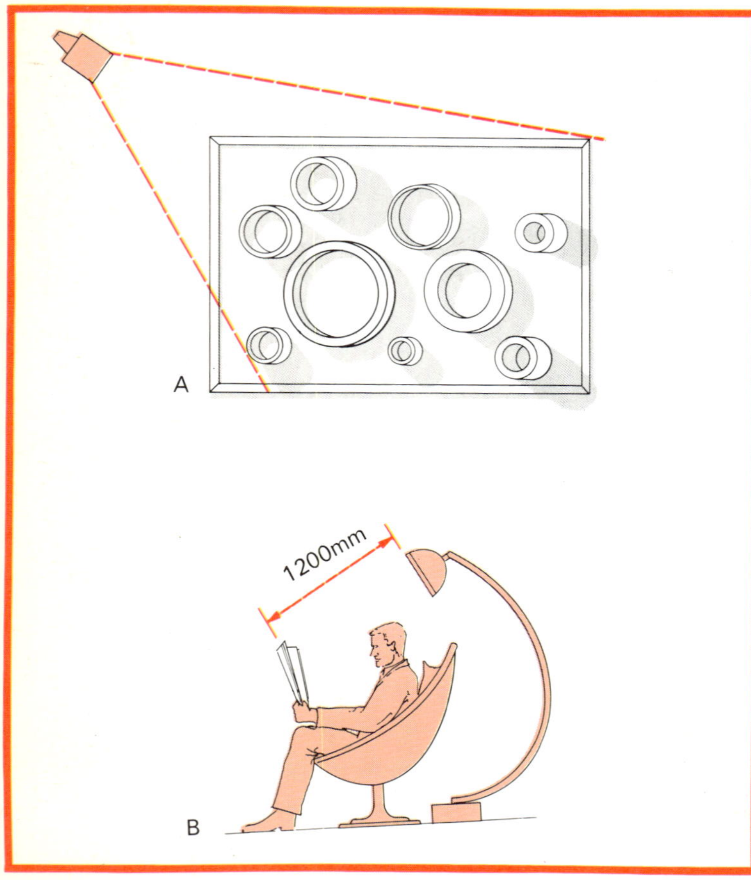

TRI-ART

DAILY TELEGRAPH

Left. *This attractively original standard lamp casts light onto the floor—avoiding the usual dazzling effect of all white decor.*

choose a dimmer that is rated for the maximum power of all the lights it controls. You must also make sure you buy the right kind of dimmer for your lights: some, for example, cannot be used with fluorescent lamps. Your supplier will be able to advise you.

Lighting levels and colours

Light levels can influence your moods in the same way as certain colours can. Very bright lights will cause glare and eventually make you irritable. Also bright lighting can make you excitable and in certain extreme cases have an intimidating effect. It is important that your lighting arrangements are such that they provide the right level of illumination for comfortable and relaxed living—and that the arrangements can be changed to allow for different activities.

The colour of the lighting can also have an effect. Normally you will use white lights but even these can give different effects. If you have fluorescent, for example, you may decide to fit tubes that match the warmth of tungsten lamps rather than the colder-looking daylight. And using a number of low-power tungsten bulbs gives a different look from that of a single, high-powered bulb. If you intend to use coloured bulbs you must take care. In general terms, red stimulates and green relaxes. But you will find that if you use either of them as the dominant light source the result can be less than pleasing. The stimulating effect of red can often turn into over-excitement while the relaxing quality of green could send you completely to sleep. Always use coloured bulbs to complement your room and not to dominate it.

Once you have sorted out your basic lighting needs it is possible to turn to other important uses for light in your living room as decoration in itself. These decorative effects can form an important source of background illumination and will transform the look of your living room. If you have bookshelves in your room why not install small spotlamps along the ceiling above or, alternatively, fit recessed lamps in the same position? The functional advantages of these lights are obvious but the decorative effect will be breathtaking, particularly if your books are attractively bound.

You can also use lights to illuminate pictures, tapestries or even a fine set of curtains. Pictures where pale colours predominate should be lit with a strip light placed just above the picture with the bulb away from the wall. In the case of darker pictures fit two or three small spotlamps just above the picture with the light directed onto it. For tapestries or curtains these two methods are unsuitable as direct light will make the pattern appear flat and dull. Place a small spotlamp just above and slightly to the side of one of the top corners of the tapestry. The light shining from this angle and onto the fabric will bring out the texture and intricacy of the pattern. (see diagram **A**.)

It is very important to achieve correct lighting for your living room. The effect on the decor of your room and your own comfort will make the fairly modest effort and expense well worth while.

HEIDEDE CARSTENSEN/STUDIO DIE WOHNFORM

Bottom left. *Pools of extra light for reading or sewing can be provided by a suitable table lamp.* ***Bottom.*** *The globe shaped lampshade provides diffused general light in a large room.*

RICHARD EINZIG

HEIDEDE CARSTENSEN/STUDIO DIE WOHNFORM

__Above.__ In this bedroom the bed is lit by fluorescent strips, fitted into recesses on either side of the bed. Extra illumination is provided by the two table lamps.

Bedroom/bathroom lighting

The importance of good lighting in the home cannot be overestimated. This is as true in your bedroom and bathroom as it is anywhere else in your home. Adequate and well planned lighting is essential for comfortable living, and will also greatly improve the appearance of your furnishings and decor.

All too often bedrooms are regarded as merely places for sleeping in—and are furnished and lit as such. However, bedrooms can double as valuable extra living areas. This is an advantage in any home, but if you have children, and live in a small house, the bedrooms can become essential for extra space.

The bedroom

Apart from providing adequate overall illumination you should install light fittings in particular areas of the room—near wardrobes and closets, over mirrors, near an easy chair, and a reading light over the bed.

Once you have decided what lighting is appropriate for your particular bedroom, you can go on to choose from the almost endless array of light fittings on the market. To do this successfully, you should list the functions that each individual fitting has to satisfy. Having done this, choose the fitting that will meet your requirements best. Don't go by appearances—a fitting which catches your eye in the shop may well prove totally unsuitable for its intended purpose. Amongst the range of fittings available you should be able to find one that both meets your requirements and satisfies your individual tastes. Remember, the intended function is the only really important consideration when choosing light fittings. For the most part, a sensible choice of shade can solve the problem of combining practicality with beauty.

The basic principles of planned lighting for living rooms are also applicable when planning lighting for bedrooms and bathrooms. Consider them in your plan. Firstly it is important that you attempt to combine the need for satisfactory overall light with the demand for localized light for particular areas of the room. If a successful combination is to be achieved, you should make sure that the total brightness of the general light source is not too great. This is important because if this light is too powerful you will not, comfortably, be able to use any of the auxiliary lights with it.

The best solution to this is to fit a dimmer switch, in place of the ordinary light switch, to operate the main light. There are two types of dimmers commonly available. The first consists of a lever and slot mechanism. As you move the lever from left to right, along the slot, light intensity increases. When the knob is furthest to the right of the slot the light is at full power. The second type is controlled by a knob, which, by turning in a clockwise direction, increases the brightness of the light. With this type of dimmer the choice of lighting levels you have is almost infinite. By the use of a dimmer switch you'll be able to keep the main light source at a fairly subdued level. Where extra light is needed you can use a standard lamp, table lamp or wall lights. This kind of lighting combination will prove much more restful to the eye than if you simply rely on the main light at full power.

Before going on to consider ideas for lighting bedrooms, some safety points need to be mentioned. All your light fittings and power points must be properly earthed. In Britain this

TIMSTREET PORTER/ELIZABETH WHITING

Above. *Here, the spotlight fitted above the bed gives excellent illumination for reading. An atmosphere of subdued intimacy can be achieved by fitting a dimmer switch.*

is laid down by law for new houses. Also, any fittings you have must possess double insulated flexes. You should check that old fittings comply with these safety requirements. If they don't you will have to replace them. Finally, if your house is the slightest bit damp, don't fit wall lamps.

The main light source

The main light source is intended to give an even overall light to a room. In most bedrooms this is provided by a central ceiling light, of one or more bulbs. In very large bedrooms there may be two ceiling light fittings for general illumination. Sometimes, but rarely in bedrooms, the main light source is provided by wall lights.

Don't put bulbs that are too powerful in the ceiling fittings. If the main lights are too bright they will be dazzling, and will soon prove very uncomfortable to live with. It's impossible to state what the maximum output of the main light source should be, as some people are more sensitive to light than others. However, you'll probably find that a total output of 100 watts is sufficient for an average size bedroom. Even in a very large bedroom 150 watts should be more than adequate. If the main light is controlled by a dimmer switch, it can be kept at a fairly subdued level for the most of the time. Other light fittings can be employed to give brighter light in areas of the room where it is needed.

Measure the maximum brightness of the main light source against the decor and colour scheme of the room, the size of the room, and the total power and number of other lights in the room.

As with other light fittings, there are a wide variety of fittings for the main ceiling light available. By far the most common is the single flex type—one or more bulbs hung from the ceiling by a single flex. You'll more than likely find that this is the kind of fitting used in your home. To get the best out of the single flex fitting, choose a shade that throws the light downwards to cover a wide floor area. Try not to choose a shade that is heavily patterned, or of a dark colour, as this will cut down the effectiveness of the light. The intention of a lampshade is to diffuse, and not reduce the light. With the choice of a sensible shade you will eliminate dazzle as well as greatly improving the appearance of the single flex fitting.

This basic fitting is capable of adaptation for increased flexibility. You could consider replacing the existing flex with a spool fitting. Substitute the flex for a much longer piece. This is then wound round a spool, which can either be fixed to the ceiling or inside a special housing. With the use of a spool fitting you'll be able to raise and lower the ceiling light at will. By lowering the fitting, light will be concentrated on the floor and the upper parts of the room will be darkened—achieving a relaxing and subdued atmosphere. By raising the light closer to the ceiling more overall illumination will be provided.

Apart from the single flex fittings there are a number of other types of ceiling light you can fit, and which are suitable for bedrooms. In a bedroom with a false ceiling, recessed lights look most effective. These can consist of ordinary tungsten bulbs, suspended by a flex from the original ceiling. Holes are cut into the false ceiling and a stainless steel, aluminium or enamelled surround is fixed. Recessed lamps look best if a number of them are placed at even distances over the ceiling area. The recesses should be kept fairly shallow to give a wider dispersal of light. If the recesses are too deep you'll find that the effect will be of 'columns' of light cast onto the floor, and not the even overall illumination you should be aiming at.

Another good idea for bedrooms with false ceilings is that of lighting 'panels'. Fluorescent strips are hung from the old ceiling to about 4in. (100mm) above the false ceiling. Opaque white plastic panels are fitted into the new ceiling. These can be removed for easy cleaning. A well diffused light results from the use of these panels.

If your bedroom has a low ceiling—a common feature in the modern house—you could consider flushing the flex hung ceiling light direct to the ceiling. Providing you choose an attractive and practical shade this idea will work —both from the functional and decorative points of view.

An alternative for the bedrooms with a low ceiling is to fit lighting 'tablets'. These consist of three or four fluorescent strips fixed, parallel to each other, on the ceiling. These are covered over by an opaque white plastic case. The most common shape is square, and similar in size to a paving stone—hence the term 'tablet'. However, other shapes can be obtained, including rectangular, round and hexagonal. The effect is similar to the 'panel' lighting described above.

Whether your main light source is operated by a dimmer or a conventional on/off switch, it would be a useful idea to fit a cord pull switch over the bed. This will enable you to operate the main light without going to the trouble of getting out of bed.

Other lights for the bedroom

When planning the lighting for your bedroom you should consider a few ideas of particular value in bedrooms. Apart from a cord pull switch over the bed, to operate the main light, you should install a reading light. Instead of the usual table lamp why not fit an adjustable wall light on the wall behind the bed—operated by a cord pull switch? This will, by simple adjustment, give a more versatile and comfortable light than a table lamp. Table lamps do not usually give a sufficiently powerful direct light for reading in bed. Also, table lamps are most commonly placed on a bedside cupboard or table—at completely the wrong height and distance for comfortable reading.

You should install adequate lighting near mirrors. Apart from the fact that mirrors are often placed in dark corners of the bedroom, you'll need to overcome the problem of reflected shadow. If you have a built-in dressing table unit, fix a fluorescent strip, horizontally, above the mirror. Alternatively, you could fit a strip light, running vertically, at either side of the mirror. Whatever method you choose, the fluorescent strips should be concealed behind timber baffles. This will prevent glare and will

Above. *Adjustable spotlights, like the ones shown here, are excellent in bedrooms. Versatility is increased when the lights are fixed to a portable stand.*

Below. *A bathroom needs adequate overall light. The long mirror in this spacious bathroom is lit by a fluorescent strip, which runs along the whole length of the mirror.*

PAF INTERNATIONAL

direct the light into the mirror. Light shining into a mirror will be reflected outwards—almost eliminating shadow. Should your bedroom have a washbasin and mirror, install a fluorescent strip, vertically, above the mirror. This light should also be concealed behind a timber baffle.

A good idea for wardrobes is to fit a fluorescent strip on the underside of the shelf. An on/off mechanism could be fitted onto the wardrobe door, so that when the door is closed the light automatically switches out, similar to a domestic refrigerator.

Bathroom lighting

Your bathroom lighting should be related to the lighting scheme you adopt for the bedroom. Generally speaking, the bathroom will need a fairly bright overall light to enable you to do such things as shave in safety.

If your bedroom directly adjoins the bedroom, the change in light between the two rooms should not be too dramatic. However, if, as is usually the case, you have to cross a landing to reach the bathroom from your bedroom, try to achieve a gradual increase in light between the bedroom, landing and bathroom.

The main light source in a bathroom is usually provided by a central ceiling light, or by one, or more, wall lights. Don't put a very powerful bulb in the main light fitting. Most bathrooms are quite small, and 60 watts is usually adequate for the main light source. It's far better to rely on a sensible choice of shade. One of the many opaque white glass shades available are ideal for bathrooms. With this type of shade you'll find that brightness is hardly reduced at all. At the same time the light is well diffused to avoid glare. Also, careful use of light coloured decor and reflective surfaces—like tiles and mirrors—will help you to achieve the maximum brightness possible from a small bulb. Apart from conventional main light fittings, there is a bathroom light available which has a built-in heating element.

The bathroom mirror, over the washbasin, should be lit by fluorescent bulbs in the way described above. If you have a lot of toiletries, in attractive bottles and packages, you could consider making use of them in an illuminated shelf display in the corner of the bathroom. This simply consists of fluorescent bulbs, concealed behind baffles, fixed under each shelf—and lighting the shelf below. Another good idea is to fix a small strip light inside the bathroom cabinet—operated by an on/off door switch.

For safety reasons the main light in your bathroom must be operated by a cord pull switch. An ordinary light switch can only be used if it is situated outside the bathroom door. If the switch is placed outside, there will have to be a glass panel over the door so that it can easily be seen, whether the bathroom is occupied or not. This acts as a safeguard against the light being turned off when someone is in the bathroom. Power points are dangerous in a bathroom, so any wall lights will need to be wired into the wall itself. As with the main light, all accessory lighting should be operated by cord pull switches.

Take a look at the lighting in your bedroom and bathroom—a small modification could dramatically improve your visual comfort.

SPACEFITTA BY LIMELIGHT

BRIAN MORRIS

Lighting for dark rooms

Dark, dreary rooms need not be a problem if you decorate them with imagination. Use colour as your main ally, together with well planned lighting and mirrors, to conjure up a feeling of more light. Extra windows, naturally, are the best way of letting in more light to dark rooms, and you may be able to afford to alter or enlarge the windows in your house. This is more of an upheaval than ingenious decorative treatments, however, and entails structural alterations.

Plan what you are going to do. If one of the rooms in your house is dark, and you want to brighten it up, don't simply increase the wattage in all the light bulbs and paint the walls and ceiling white. There are more subtle—and more effective—treatments for dreary areas, and each room must be dealt with according to its function, shape and size.

Using colour

You can obtain colour charts specially planned to help you create light colour schemes for dark rooms. Different rooms need different treatments, so consider what effect you want to achieve before you go ahead. For instance, an all-white scheme in a living room can make it seem light and airy, but an all-white bathroom would be cold and clinical without the addition of bright pictures, towels and other accessories.

Choose cheerful colours like daffodil yellow for tall, dark rooms, or brighten them up with a gay wallpaper, paint the ceiling in one of the paler colours of the wallpaper and use masses of white on the woodwork and furniture. Low, dark rooms often look best with a white ceiling, but if you want to paint it the same colour as the walls, you can avoid a lowering effect by picking out the cornice in brilliant white, and add bright-looking furniture in white, sky blue or yellow.

Using bold colours in a dark room needs special care. Bright red in a small hall looks warm and welcoming, but it can also look dark unless it is mixed with a lot of white plus a big mirror and strong lighting. Remember that pale colours like white, grey and pastel blue are space givers, whereas more dense colours like emerald green and scarlet or orange advance, making a room seem smaller.

In the northern hemisphere, north- and east-facing rooms need warm colours to help them feel sunny, so yellows and oranges are good dominant colours. South- and west-facing rooms will be sunny anyway, even if they are dark, so you can safely use cooler colours like white, pale blue or very pale green.

Heavy colours like black and deep brown should be used sparingly in a dark room, although one wall covered in chocolate brown hessian will give a warm feel to a room. A black-covered kitchen bench could be too sombre in a dark kitchen, but if you choose orange, blue, or even white, the worktop will act as a reflective surface, making the dull room lighter.

The importance of combining colour planning and planned lighting cannot be overstressed.

Window treatments

Enlarging windows is the most positive way of creating more light in dark rooms. If you don't want to widen your windows, a less major operation is to lengthen them to floor level and put in French windows in the basement and ground floor rooms. Long windows can also be put in bedrooms, but this is not a good idea in a child's room. A small fake balcony outside, attached to a strong pair of brackets, will make the window look more natural; this could take the form of a wrought iron railing, or a big painted brass fender with room for a trough of flowers inside.

To catch more light in a ground floor room, change a plain window for a bow-shaped one, or a larger bay window. Installing a bay window would mean extending the walls outwards to take in the shape of the bay. In Britain, any major alteration of this kind would mean applying for planning and/or building permission. Check on the details of installation of bow windows before beginning the project and double check the details of building regulations regarding minimum window area in relation to floor area.

You can add more light fairly simply by changing the type of window inside the original frame. Small-paned or leaded windows can be replaced by ones with larger panes, but remember that the glass will be more expensive to replace if it gets broken. Many Edwardian houses still have coloured or stained glass panes, particularly in hall or living room windows. If you swap these for plain glass, more light is easily let in.

Remember that altering a window will change the whole appearance of a house from the outside. Well thought-out enlarged windows will add a feeling of light both inside and outside the house. But windows that do not blend with the design of the house, especially if it is a decidedly period one, can ruin an otherwise attractive front, although they may improve matters inside.

Curtains and blinds

The main rule to remember in a dark room is never to use dark curtains or blinds. Light window coverings will help reflect what daylight there is in the room. Don't half-draw curtains; instead, make the rail long enough so that the curtains can be pulled back to clear the edge of the window frame, thus letting in as much light as possible.

Blinds are particularly good in dark rooms because they can be neatly rolled up in the daytime so that they don't obscure the light at all. Shutters are another good choice, because they fold neatly out of the way. In a dark bathroom, you can make your own shutters out of figured hardboard, painted white, or use white-painted louvred shutters. They can be left closed during the day for privacy and will still let in a certain amount of light which can be supplemented by a concealed tungsten strip light behind them if necessary.

It is best to keep the feeling of light coming through a window in a dark room by painting the frame white, both inside and out. If you need net curtains, choose a crisp white rather than a pastel shade; the heavy fish net type let in more light than the more dense plain ones. A pretty effect can be achieved by using a thick Nottingham lace with a fairly open pattern to let in the light.

Lighting

Well-planned lighting can transform a dark room, and there are several tricks to remember if you want to make the most of it. One is to put table or standard lights near the window, even on either side of it. Placed in this way, they will increase the feeling of light coming through the window.

Hidden lighting behind a curtain pelmet is another useful trick. This gives a good background light, helps to frame the window, reflects a light-coloured ceiling, and again will help to increase the feeling of daylight in a room. Strip lights used in this way, or fitted behind a false cornice, should be tungsten tubes, which give a more natural light than fluorescent tubes. Main lighting over the rest of the room can come from table lamps.

One or two spotlights reflected on to the ceiling will help brighten up a dark room. Highlight a dark corner with an eyeball spot let into the ceiling (expensive to install, but worth it) and add to the effect with a huge jug of pale grasses or gay flowers.

In a dark hall, living room or kitchen, use lighting hidden above a suspended ceiling (see page 44). Hidden strip lights above a work surface help in a dark kitchen, but more light will be needed at ceiling level for general illumination.

Rooflights

Rooflights or skylights are invaluable for giving extra light, particularly at the top of a stair well, or let into a flat or sloping roof over a dark room. If you do fit one into a flat roof, you will need to alter the positions of the joists to make room for it—a square one is the easiest.

***Left.** Make a dark room seem much more bright and cheerful through the use of mirrors, mirrortiles and other shiny surfaces such as in this unusual bathroom.*

HEIDEDE CARSTENSEN/STUDIO DIE WOHNFORM

ELIZABETH WHITING

HEIDEDE CARSTENSEN/STUDIO DIE WOHNFORM

HEIDEDE CARSTENSEN/STUDIO DIE WOHNFORM

MICHAEL BOYS

One of the problems which arise with rooflights is that they let in water if they have not been fitted properly. In order to make your rooflight watertight, fit a cowling with a lip round to keep out the rain. See page 83 for detailed information on choosing and installing rooflights.

Mirrors

A well-placed mirror is another great help in a dark or gloomy room. The best method is to hang a large mirror on the wall opposite the windows so that it reflects the light coming into the room, doubling its effect. Just as a light near a window will increase light, so will a table light placed near a mirror. The mirror reflects the light again and again, creating an impression of much more light.

Full length mirrors transform a dark hall, passage or landing, particularly if they are placed so that they reflect bright, sunny furnishings. In a minute, dark bathroom, you can go further and have a complete wall of mirrors. Add a strip light concealed behind a fake pelmet above the mirror, casting light down where it is needed.

Look for huge old mirrors in junk shops and use them as bedheads in dark bedrooms. If you cover all the doors of a fitted row of cupboards with mirrors, it will help to create a feeling of more light and space. 'The magic thats all done with mirrors' has more ideas.

Dark areas

Basement rooms or flats can be brightened up considerably by decorating the area outside in light colours. The simplest and cheapest way is to paint everything, brick walls, steps, doors and dustbin cupboard in a light colour such as white or yellow. Dark-leaved climbing plants will darken an area, so choose the type with variegated leaves if you want greenery outside the window. Plant white-painted tubs with bright flowers like marigolds and nasturtiums and arrange them on the outside window sill or at the base of the wall if you have full-length windows.

Mirrors can be used here, too. Fix a mirror on the area wall opposite the window; the bigger the mirror the better. Any old cracked or damaged mirror will do, and you can arrange a 'frame' of plants around it to make the area seem larger. If you don't want to have to look after plants all the time, try painting a bright flowery mural on the area wall opposite the window. Even a simple yellow sun on a blue background looks effective, and makes a light-hearted treatment for a dull area.

Top left. *The long mirror over the fire increases the apparent depth of the room and adds a welcoming glow to the decor.*
Bottom left. *A dark interior transformed into an attractive, light room through the use of light colours and bright surfaces.*
Top right. *Slatted roof lights and delicately coloured glass panels contrive to lighten this dark internal corridor.*
Middle picture, right. *Mirrors give the illusion of doubling the space, but they can be too glaring. Here the effect is counteracted by the restful green decor.*
Bottom right. *A rooflight of reinforced glass provides burglar-proof light relief.*

Ideas room by room

Hall and stairs. You can increase the light cast over a dark hall and staircase by letting in a fan light or small window above the front door. You can create a feeling of light and depth by papering the ceiling in the hall with a bright, geometric paper. Add plenty of white, particularly on the woodwork, plus a mirror to reflect all the light things. Borrow light from the rooms leading off the hall and landing by fixing glass panels in doors, or by installing 'windows' in the walls. These can be disguised as shelving alcoves inside the rooms, and help to let more light onto the stairs. Remember the safety aspect of lighting for staircases and landings, it must be adequate as well as welcoming and bright.

White-painted banisters will look less heavy and dark than natural wood, and an open staircase with a light carpet will give a brighter, more spacious feeling. Don't obscure the window on the landing, but help it by fixing hidden lights behind plants or ornaments strategically placed to highlight corners.

Living rooms. Large windows, combined with fresh-looking curtains or blinds, warm lighting and a pleasant colour scheme all help to emphasize brightness in otherwise dark rooms. Be careful in your choice of furniture, however, as big heavy pieces like a mahogany desk or dark leather-covered sofa can be too overpowering if you want light. Don't mix too many different colours together, as this looks messy; a plain colour will usually create a lighter feeling than a pattern, particularly where carpets are concerned.

You can create a much brighter appearance simply by painting the floorboards white to reflect the ceiling, then adding shaggy mats, Spanish rugs or rush matting to make it more practical. Use plenty of table lamps with white shades to create the maximum amount of light, with spotlights in dark corners, and concealed lights in alcoves or behind curtains as background lighting.

Dining rooms. The dining room is often the darkest room in the house, and if you tend to use it only in the evening, this does not matter too much, as then you can rely entirely on artificial lighting. If you have a young family, a dining room needs to be a light and airy place, and often a one-colour treatment is the most effective, with the walls, ceiling, floor and furniture painted in the same gay colour. At night the appearance of the room can be changed dramatically by the addition of a patterned tablecloth with matching blinds at the windows, and softer lighting. A low hanging light over the dining table will focus attention on it. If you think blinds are too harsh, hang fake curtains at each side of the window; these will frame the blind, but will not obscure the light by day.

Kitchens. As well as a gay colour scheme and wisely chosen lighting, you should look for accessories like tiles—for worktops or floor, storage jars and pretty pots and pans which will add brightness to your kitchen. However, you must remember that while a plain flooring in a pale colour like white or yellow may reflect extra light, it will need a lot of cleaning if it is not to look tatty. Instead of a dark flooring, however, go for an attractively patterned sheet vinyl, which will look bright but not show every speck of dirt.

One of the best ways of lighting a low kitchen is to fit downlighters into the ceiling. They don't encroach on the headroom, but are particularly efficient if they are placed strategically above worktops, cooker and sink. Another way of letting in more light is to fit a larger than average serving hatch, thus borrowing light from the adjoining room. In a kitchen, gloss-painted walls combine extra reflected brightness with ease of cleaning.

Bedrooms. Bedrooms should feel warm, so a pale colour is usually best avoided, even for a dark room. A lot of white in a north-facing bedroom is a particularly bad idea, and it is better to choose a sunnier colour. If you want the impression created by white, choose beige or oatmeal, which give a light feeling without looking cold. You can create an attractive effect by using a patterned wallpaper, preferably on a white background. Make up the curtains and bedspread in the same fabric, if possible to match the wallpaper, and pick out all the woodwork in white paint. The ceiling can be painted in one of the paler colours of the wallpaper, with a matching carpet. Carpets are expensive, though, and a cheaper way of brightening the floor is to paint the boards in a light colour to match fabric or wallpaper used elsewhere, whether it is yellow, sky blue or pink. It is possible to paint linoleum in a matching shade.

Bathrooms and lavatories. Oddly enough, one of the best ways of dealing with a dark bathroom or lavatory is to decorate it in deep vibrant colours. These will help to make the room seem warmer, and when combined with plenty of strong lighting it will become much brighter. Since you are often scantily clad when you are in the bathroom, it is important that it is a cosy room, so try to avoid the traditional pastel shades of pink, blue and green if your bathroom tends to be cold. Here a deep chocolate brown with bright towels and blind will create a more inviting feeling.

Lavatory windows usually have frosted glass in them, so the view out is not important. A permanently-closed blind which matches the wallpaper will give a warmer feel, and a bright, low light hanging from a dark painted ceiling will minimize the lofty feeling of a very tall room. You can achieve an unusual effect very cheaply by papering the walls of a lavatory or bathroom with newspaper. Paint one or two coats of clear wallpaper lacquer on top to give a washable surface, and you have a bright black and white room.

Obscured glass sometimes keeps out more daylight than net curtains, so check whether you really need frosted glass in your bathroom and lavatory windows. If you are overlooked, you may find that only the lower part of the window needs to be obscured, and that the top can be fitted with clear glass.

Don't resign yourself to having dark, dreary rooms. Use the ideas given here to transform the dull parts of your house into bright, welcoming rooms.

Fluorescent lighting

While the possibilities and various applications of fluorescent lighting around the home are becoming more widely appreciated as its technical quality increases, it is still a subject which is not very well understood in entirety. This chapter outlines the principles behind its operation, and explains the technology involved in common domestic forms of it. This should enable you to choose the type of fluorescent tube most suitable for the situation at hand, and come to an understanding of the reasons underlying the guidelines suggested.

Experience of the early types of fluorescent lighting may have induced some 'consumer resistance', but more recent developments have resulted in improvements, especially in the important area of colour rendering. This is the effect of this form of lighting on the appearance of the objects illuminated. At one stage of its development, fluorescent lighting lent a rather unnatural air to the objects and occupants of a room. Fluorescent tubes now, however, rival tungsten-filament lighting in this respect. Although this has resulted in some loss of efficiency, the fluorescent lamp is still very much superior in this respect and, with its longer life, it is still a better economic proposition despite the higher initial cost. However, the cost of illuminating the average household is comparatively small, so you may not consider the economic advantage to be a major factor.

Principles of operation

The most common type of fluorescent lamp is the 'hot-cathode' type, consisting of a glass tube containing argon gas and a small amount of mercury at a very low pressure. Electrodes made of tungsten wire, coated with barium or strontium oxides for high electron emission (the electron is one of the elementary particles found in all atoms, and the basis of electricity), are sealed into each end of the tube. Finally, the inner surface of the tube is coated with a fluorescent powder or 'phosphor' (see Fig.1).

External control gear (described below) provides a cathode heating current, an initial high voltage to start a discharge (i.e. ionization of the gas between the electrodes), and some form of current limiter to prevent excessive current once a discharge has been initiated. (Note: with an A.C. supply, which is almost invariably used to operate fluorescent tubes, each electrode behaves alternatively as an anode and a cathode as the polarity or direction of the current changes.)

Electrons emitted by the heated cathode are accelerated by the high voltage on the anode (which attracts them), causing a discharge through the argon which rapidly gives rise to ionization of the mercury vapour. This ionization produces very little visible light, but it does produce the ultraviolet radiation which excites the

Right. *A fluorescent tube used to provide strip lighting for a kitchen working surface. The improved 'colour rendering' properties of the modern tube are of great benefit here.*

fluorescent 'phosphor' coating (i.e. the coating effectively takes in one sort of light and converts it into another sort) It is this excitation which produces visible light. The colour content of the light is largely dependent upon the types of phosphors used in the fluorescent coating.

Starter circuits

There are many different types of starter circuits for fluorescent tubes, but those most likely to be encountered are the glow starter, the thermal starter and the quick-start type. The first two types employ a 'starter switch', but the quick-start method avoids the necessity for one. Probably the most common type, particularly in domestic use, is the glow starter, shown in Fig.2. The starter switch consists of two electrodes, one or both of which is bimetallic (i.e. is made of two different metals), sealed into a small glass bulb containing an inert gas like argon or a mixture of inert gases. The glass bulb is normally housed in a cylindrical metal canister with a small capacitor (C1). Current for the tube is drawn from the mains at L and N (for 'live' and 'neutral'), the usual wall on/off switch being in the L conductor, as for a traditional filament lamp. The capacitor C2 may be ignored for the moment.

When you switch the light on, current flows through the choke (an inductor component) and the heaters to the glow switch, causing a discharge between the glow switch contacts. (The choke has turns of wire on a core: it provides a high voltage surge, and also limits the current flowing, once the tube has 'struck'.) This heats the switch strips so that they bend slightly and touch, completing the circuit. This allows a larger current to flow through the tube heaters, raising their temperature. At the same time the discharge in the glow switch has ceased because its contacts have closed. The contact strips therefore cool until they separate. This interruption of the circuit causes a voltage surge which 'strikes' or starts the tube. Voltage across the glow switch is now too low for it to heat, so it remains open and the tube, once started, remains lit. The capacitor C1 is to reduce radio interference from the circuit. Due to the choke, current taken by the circuit lags behind voltage, that is, one is out of phase with the other. This results in a poor 'power factor'. Capacitor C2 helps to remedy this, and so it is called the 'power factor capacitor'.

The thermal starter circuit, which is more likely to be found in industrial installations, is illustrated in Fig.3. This type also depends upon a starter switch with bimetallic contacts. When the mains supply is switched on, current passes via the closed contacts of the switch to the cathode heaters and to a small heater coil in the starter switch can. After a few seconds, the switch contacts open, causing a high voltage surge which causes the tube to 'strike'. The tube current passing through the starter heater is sufficient to keep the switch contacts apart.

One of a number of quick-start circuits is shown in Fig.4. The principal differences between this type and the two just described are the absence of a starter switch, and the presence of an earthed strip running the length of the tube. This strip can be the chassis or frame supporting the tube, or the tube itself can have a metallic conducting strip cemented to its outside. The high potential difference between the earthed strip and the tube electrodes helps to initiate ionization. In some types of quick-start circuit, the transformer steps up the supply voltage to the tube as an extra aid to good starting. Fitting this type of fluorescent lighting in your home may mean that, if your house was wired some time ago, a separate earth wire has to be run from the fitting to the main earthing point in your house. Examine your wiring carefully before you begin.

Supply voltages

The vast majority of fluorescent tubes are operated from the normal A.C. mains supply of between 220 and 240 volts, at a frequency of 50Hz (1 Hertz or Hz means 1 cycle per second). Although it is possible, with certain precautions,

ELECTRICITY COUNCIL

Left. *Small fluorescent tubes illuminating an ornamental alcove. Many modern tubes have 'diffuser' and 'baffle' fitting attachments to assist in the even distribution of light.*

to run fluorescent tubes from D.C. mains supplies, this method of operation is not discussed here as it is most unlikely that this type of supply will be encountered. An important recent development is a range of fluorescent lamp fittings for use with 12 or 24 volt D.C. supplies. The incorporated control gear includes a transistorised circuit which converts the D.C. supply into a high voltage, high frequency A.C. supply for the tube, which at present may be either 8 or 13 watt. These are extremely useful, as they may be run from a car battery with a much lower current than would be required for a tungsten-filament lamp of an equivalent light output. For example, the 12 volt, 8 watt tube gives a light output equivalent to a 40 watt tungsten-filament lamp, but a current of only 0.8 amp is required from the supply. Because a high frequency supply is used, there is no flicker or 'hum', which may occur when running from a 50 Hertz mains supply, although these effects are not usually noticeable.

Tube sizes and ratings

Before deciding to fit a fluorescent light, you should consider the sizes and types available, the size of the room or area to be lit, and the wattage that will be necessary. You should also consider if a coloured tube is required. The tube sizes and ratings in common use are shown in Table 1. Some of these tubes may not be available in those 'colours' most suitable for domestic use (also see Table 3: 'Colour Rendering').

Comparison of light outputs

It is useful to make a comparison of the light outputs (which are measured in 'lumens') between ordinary tungsten and fluorescent lights. Table 2 provides a comparison between the initial light output in lumens for 'White' fluorescent tubes and for normal tungsten-filament single-coil light bulbs worked off standard 240 volt supply. You should bear in mind that the term 'White' used in this connection is one of the categories adopted for fluorescent tubes, to provide an indication of colour rendering properties. It is one of the most 'efficient' (i.e. use of power is economical) types, providing as much light output or more than the other 'colours'. Comparative figures of light output efficiency for different 'colours' are given in Table 3, taking 'White' as the base reference (i.e. this is the most efficient).

Although efficiencies vary according to wattage and sizes, as a general rule the light output of a fluorescent tube is about five times that of a tungsten-filament lamp of the same wattage (i.e. power consumption). Therefore, if the same output as from a tungsten lamp is required, the wattage you choose may be proportionately lower. In practice, the extra efficiency of the fluorescent is such that you can afford to forego some efficiency to obtain a better performance in another aspect, such as colour rendering. Tables 2 and 3 show that the least efficient basic 'White' tube, which you might choose for its colour rendering properties,

TABLE 1: Available sizes and ratings of fluorescent straight (1A) and circular (1B) tubes.

1A	Length of tube	150mm (6in.)	225mm (9in.)	300mm (12in.)	450mm (18in.)	525mm (21in.)	600mm (2ft)	900mm (3ft)	1.2m (4ft)	1.5m (5ft)
	Wattage of 16mm ($\frac{5}{8}$in.) dia. tube	4 watts	6 watts	8 watts	—	13 watts	—	—	—	—
	Wattage of 26mm (1in.) dia. tube	—	—	—	15 watts	—	—	30 watts	—	50 watts
	Wattage of 38mm ($1\frac{1}{2}$in.) dia. tube	—	—	—	15 watts	—	20 & 40 watts	30 watts	40 watts	65 & 80 watts

1B	Overall diameter	400mm (16in.)	400mm (16in.)	300mm (12in.)	200mm (8in.)
	Wattage rating	60 watts	40 watts	32 watts	22 watts

TABLE 2: Comparison of light outputs from fluorescent (2A) and tungsten-filament (2B) lighting. Imperial equivalents to lengths and diameters are approximate.

2A	Rating	125 watts	85 watts	50 watts	40 watts	30 watts	20 watts	15 watts
	Length	2.4m (8ft)	2.4m (8ft)	1.5m (5ft)	1.2m (4ft)	900mm (3ft)	600mm (2ft)	450mm (1.5ft)
	Diameter	38mm (1.5in.)	38mm (1.5in.)	26mm (1in.)	38mm (1.5in.)	38mm (1.5in.)	38mm (1.5in.)	26mm (1in.)
	Output	9,400 lumens	7,300 lumens	3,700 lumens	3,000 lumens	2,100 lumens	1,200 lumens	950 lumens
2B	Rating	500 watts	300 watts	200 watts	150 watts	100 watts	60 watts	40 watts
	Type	clear	clear	clear	frosted	frosted	frosted	frosted
	Output	8,200 lumens	4,550 lumens	2,880 lumens	2,060 lumens	1,230 lumens	610 lumens	340 lumens

TABLE 3: Colour rendering properties of various types of fluorescent lighting on the market at present.

Description	Relative efficiency %	Colour rendering properties
White	100	These tubes emphasize yellow and green and are rather poor on reds. The Warm White is slightly better on colours and is useful when high output is required.
Daylight	94	
Warm White	98	
De Luxe Warm White	66	Good colour rendering; produces a light similar to tungsten-filament and may be used anywhere in the home.
Natural	70	Reasonable colour rendering, but slightly 'cold' appearance.
De Luxe Natural	49	Good on all colours although red is slightly emphasized. Kind to complexions. Relatively inefficient but is still two or three times more efficient than tungsten-filament.

is still much more efficient (in terms of light output) than ordinary tungsten filament lamps.

The life expectancy for a fluorescent tube is around five thousand hours, compared to one thousand hours for the tungsten-filament lamp, although in practice these figures may be exceeded for either type. Again, this is another economic factor in favour of the fluorescent tube, which goes some way to offsetting the higher initial cost. Frequent switching on and off considerably reduces the life of both types and will also reduce the life of the fluorescent tube starters.

Colour rendering

The colour rendering properties of domestic fluorescent tubes are summarized in Table 3, which also indicates the efficiency in terms of output ('lumens') relative to the standard 'White' tube. There may be some variation in colour rendering properties of tubes of the same name made by different manufacturers.

Some coloured tubes are also manufactured —red, blue, pink, and peach amongst others. These tubes are mainly intended for decorative purposes, particularly in commercial applications, although pink and peach tubes may be available for domestic use. All coloured types have poor colour rendering properties and are generally inefficient relative to the 'White' types.

Assessing applications

A variety of straight and circular fluorescent tube fittings is made for use in the home, either for ceiling or wall mounting. The necessary starting gear is supplied with the fitting. For general illumination purposes, where colour rendering is not of major importance, such as kitchens, bathrooms, garages or work rooms, the fluorescent tube is superior to any other type in light distribution and running costs. Straight tubes are normally used in these areas, generally either 3ft (30 watt) or 4ft (40 watt). A 'Warm White' tube is the popular choice here.

For supplementary lighting, such as that for dressing-tables, bedlights, mirrors, and curtains, the fluorescent tube does not have such great advantages over tungsten-filament strip lighting. If colour rendering is important, you may need to use the better type of tube (see Table 3). However, for a particular light output, the fluorescent tube is very much cooler. Do bear in mind that this is normally because a more efficient, lower wattage fluorescent can be utilized in place of filament lighting. A fluorescent lamp, nevertheless, does give off a fair amount of heat, as approximately 80% of the input power to a fluorescent lamp is wasted in heat losses. Take this into account if a tube is boxed in or if a wooden baffle is used, as warping due to the heat may occur. You must allow sufficient space for cooling air to circulate and dissipate the heat.

Attractive wall-mounted fluorescent fittings with a range of baffles and diffusers to suit particular purposes are now available. A 3ft

Fig. 1

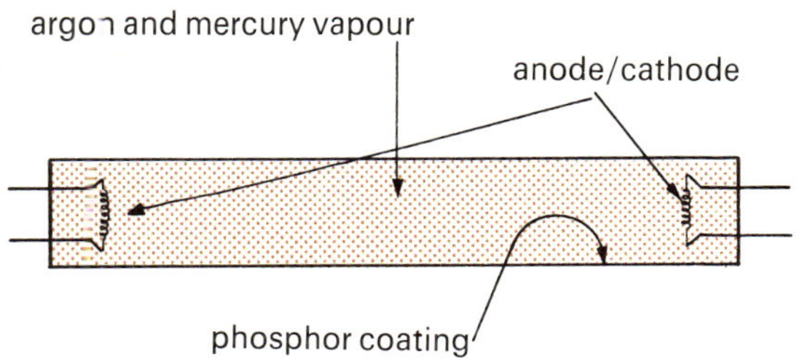

Fig. 2

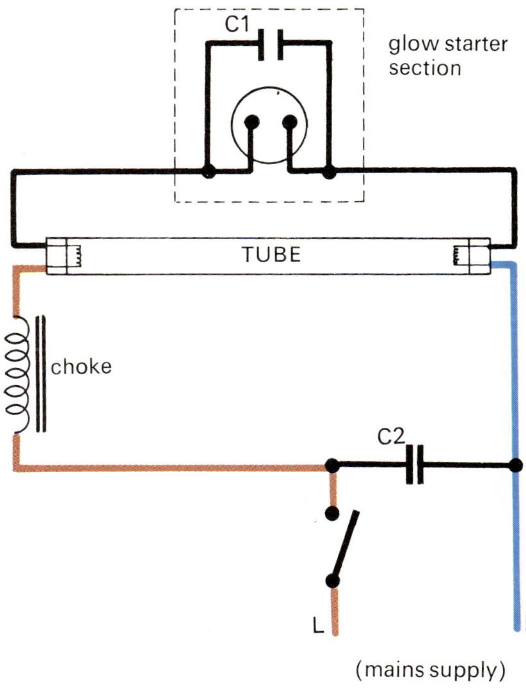

Fig.1 (above). *A schematic outline of the fluorescent tube's basic composition. The ionization induced in the gases between the electrodes (each of which is the anode and the cathode in turn, as the polarity of the applied A.C. mains voltage changes through each cycle) is the physical principle behind its operation. The phosphor coating absorbs this primary radiation, then re-emits it as visible (i.e. 'illuminating') light.*
Fig.2 (right). *A simplified circuit diagram for the fluorescent tube utilizing the 'glow starter'.*
Fig.3 (below). *A schematic diagram for the 'thermal starter' circuitry.*
Fig.4 (below right). *A schematic diagram for the circuitry to one type of 'quick start' tube.*

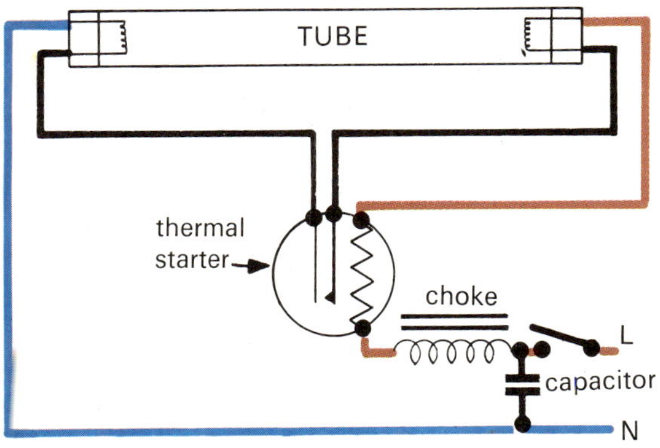

Fig. 3

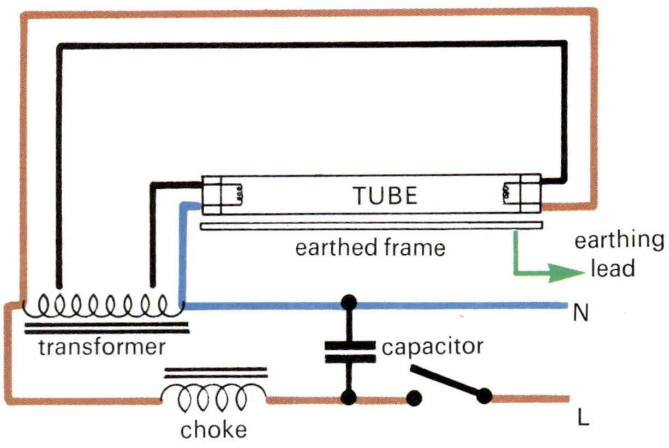

Fig. 4

(30 watt) straight tube is most commonly used, but smaller types are useful for high lighting displays etc. One example is an 18in. (15 watt) fitting, which incorporates a shuttered electric razor socket. A special version is available for use in bathrooms, to which more stringent safety regulations apply.

You may use circular tubes with glass diffuser discs to replace ordinary filament lamps. The whole fitting, including starting gear, is simply screwed into the position formerly occupied by the ceiling rose. Often a 16in. (40 or 60 watt) tube will satisfy your needs. The most suitable place for this type is probably a hall or landing, although they are sometimes used in kitchens. If a dimmer switch is to be used, it must be of a type specially designed for operation with fluorescent tubes, because of the fluorescent control gear used.

Precautions

Great care should be used when handling fluorescent tubes to avoid breakage. The very low pressure in the tubes will cause an implosion which may be severe, particularly with the larger tubes. In addition to this, the mercury and other chemicals in the tube could cause poisoning. This is especially the case if the broken glass causes a cut through which the chemicals may enter the bloodstream, or if they are inhaled. You should therefore be very careful when discarding spent tubes.

Although ultra-violet radiation is produced when the tubes are in operation, virtually all of this radiation is absorbed by the fluorescent coating and the glass tube. Radio waves are also produced which, although not harmful, may cause interference to radio or television sets if the tube is close to the aerial.

Above. *In this instance, a long fluorescent tube and its chassis have been slotted into a recess in the ceiling.*
Below. *Two tubes of different lengths have been turned to two distinct uses here—one as a reading lamp, the other as a dresser light.*

The rapid switching of the tube at a frequency of 100 Hertz (i.e. each electrode is both anode and cathode in one cycle) due to the alternating mains supply is not normally visible because of the 'persistence of vision', but it may give rise to stroboscopic effects when moving parts are viewed under fluorescent lighting. Moving parts may appear to be moving only slowly or to be stationary. The trouble is mainly confined to industrial applications, and special twin-tube circuits have been designed to overcome the problem. You are unlikely to encounter it in your home.

To sum up, fluorescent lighting is a good deal more complicated from the technical point of view than normal tungsten-filament lighting, but the advantages it makes available more than compensate for the effort involved.

Above. *The nodding and semi-hypnotic effect of the mare's tail fibre lamp is enhanced as the lazy transition through the colours of the spectrum is superimposed on the small random oscillations of the point sources of light at the ends of the individual fibres.*

Fibre optics-decorating with light

From time to time, inventions and ideas born out of some technical necessity in a remote branch of science are adapted for use in the home. One of the latest developments to find its way into the lounge room lies in the field of Fibre Optics, and the subdued illumination source, lightshow lamps that utilize one of these newly discovered applications of established scientific principles are only now starting to appear in the shops. As with many new ideas, though, the cost at first is more than you may be prepared to pay. By assembling your own version however, you can save up to half the store-bought cost of a Fibre Optic light—for no more than a few hours' work!

The so-called 'mare's tail' arrangement of the fibres into a lamp is the most obvious way to incorporate these new innovations into your interior design scheme, but suggestions for further uses you may like to put them to are included in this chapter.

The underlying ideas

The theory behind Fibre Optics is really quite simple—the individual fibres (of which there are thousands in the normal run of mare's tail) act as guides for the light source positioned beneath the ferrule that binds the fibres together at the base. Each fibre consists of a central length of glass sheathed in a covering of a glass that is

1

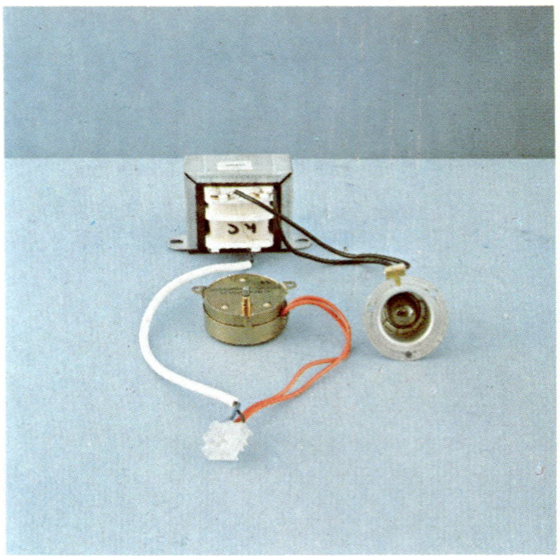

2

3

4

5

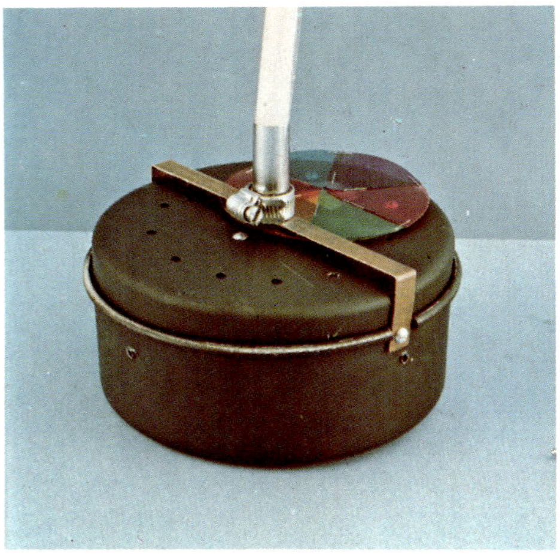

6

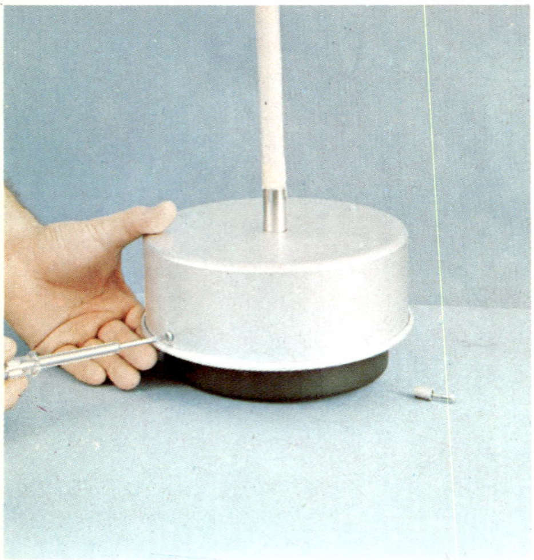

7

8

what is termed 'less optically dense'—in other words, there are two types of glass in each fibre, one running along the central axis of the fibre and the other forming an envelope around it.

When the light shines up through the central fibre a phenomenon called 'total internal reflection' takes place, because of the different optical densities of the two types of glass and the glancing angles of incidence of the rays at the interface of the two glasses. This traps the light inside the fibre and effectively channels it up to the top. Another example of this is when you look upwards from beneath the surface of a pool—the light cannot 'get out' of a large section of the surface of the water above you, and is in fact reflected back rather than transmitted through the water into the air. The result is that you can only 'see' the surface, and not the world above it, despite the fact that water is completely transparent! Fig.2 shows (schematically) what is happening to the light inside the fibre—apart from reflecting the light from side to side up the length, the main reason these fibres work so well lies in the fact that very little light 'leaks' out through the sides of the envelope glass, so the effect is not watered down before you can see it.

The mare's tail lamp

Perhaps the most simple—and certainly one of the most spectacular—of the applications of **Fibre optics in the home is the mare's tail lamp like that shown on page 39. Whether covered or uncovered, the fibres are moved gently by the** small air currents set up by the difference in temperature between the ends and the base (which holds the primary light source—in this case, a car sidelight bulb) With the addition of a colour wheel (in the base section), which is rotated by a small electric motor at around one or two revolutions per minute, the pin-points of light at the ends of the gently moving fibres slowly change colour, through as many of the colours of the spectrum as you wish to include.

A series of step-by-step photographs showing the assembly of the basic components of the mare's tail lamp is given on this page. There are not, of course, any strict design and assembly rules, and any compact and inexpensive design that you may arrive at will probably do the job just as well as this one. This design, however, does have the advantage of using materials and components that can be obtained easily and cheaply, and combines this with all-round simplicity and durability. Any component changes, modifications, or short-cuts that may occur to you should be considered in the light of the following:

—the light box, which holds all the working

Left. *Step-by-step series for the assembly of the fibre optic light box, according to the design given in this chapter:*
(1) *three-core mains flex is led through a rubber grommet into the base; only some of the holes shown are drilled for fitting the parts, the rest serving as ventilation openings;*
(2) *both transformer and motor are run off the mains, connected in at the junction block;*
(3) *the cut-down upper piece, fitted with lampholder and motor, is also well ventilated;*
(4) *with the light box parts installed, the*

Fig. 1

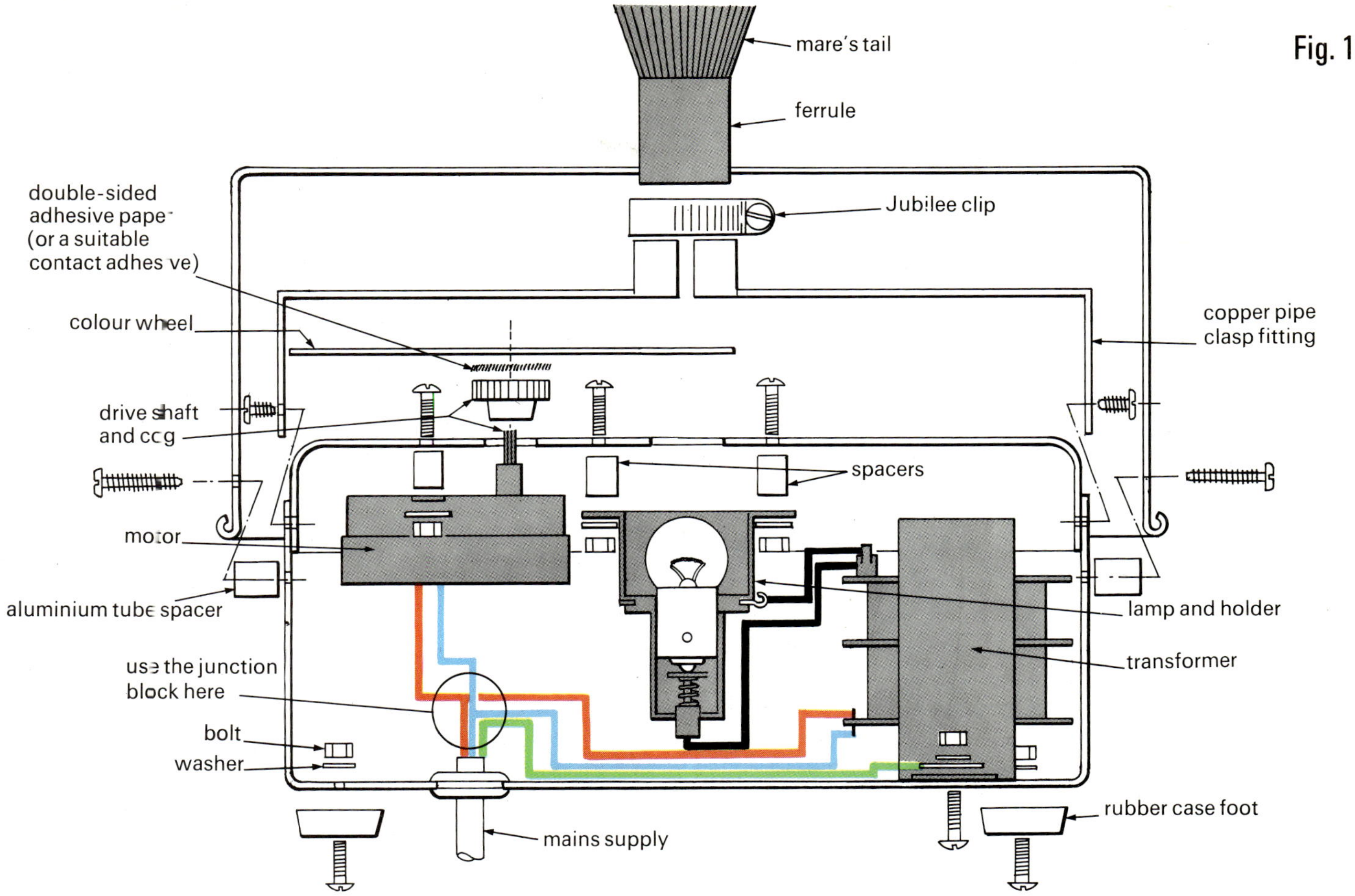

Fig. 2

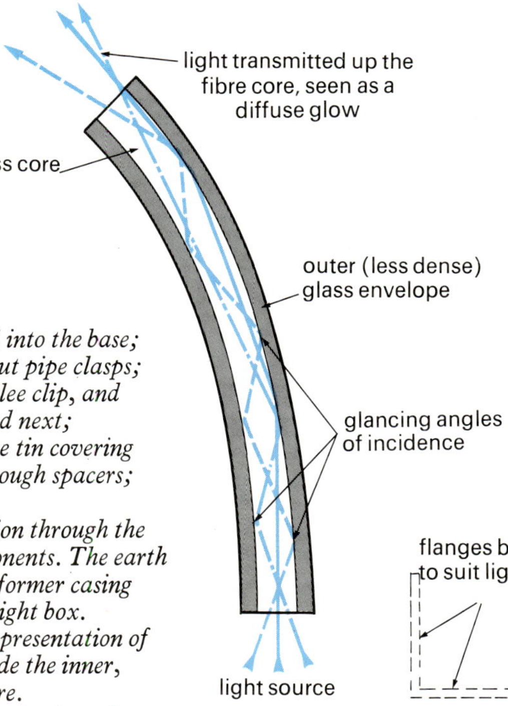

Fig. 3

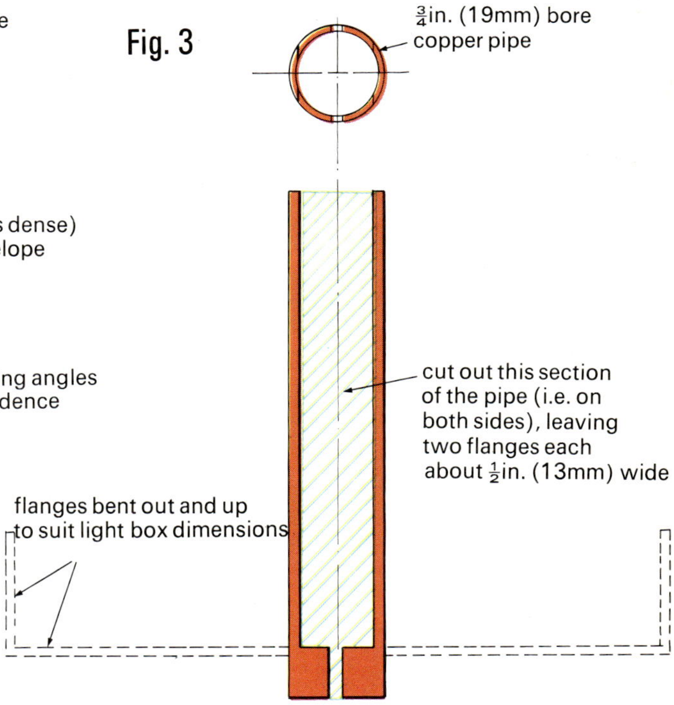

upper piece is simply inserted into the base; ***(5)*** *screwing on the splayed out pipe clasps;* ***(6)*** *the colour wheel, the Jubilee clip, and the mare's tail itself are added next;* ***(7)*** *the larger aluminium cake tin covering the light box is screwed in through spacers;* ***(8)*** *the finished product.*
Fig.1 (above). *A cross section through the lamp base, showing all components. The earth wire is connected to the transformer casing and hence to the body of the light box.*
Fig.2 (right). *Schematic representation of 'total internal reflection' inside the inner, optically denser, core of a fibre.*
Fig.3 (far right). *To make the clasps for the ferrule, some copper tube is cut down and splayed out (by careful bending) as shown.*

TRI-ART

parts as described below, must have provisions for proper ventilation. Since it is a confined area which contains a light bulb generating both heat and light, air must pass in and out to disperse the heat—otherwise the ends of the fibres bound together at the ferrule will go brown. This means that not all the light from the bulb will be able to get into the fibres, and so the total effect will be toned down. In the design given below several holes have been drilled in the light box sections, and the final, or covering part of the light box is spaced out from the other sections as it is screwed on to ensure maximum circulation of air. The wattage of the bulb is crucial here, as too much power will result in more heat being generated than can readily be transmitted through and out of the light box;
—generally speaking, the smaller you can make the light box the better. A hemisphere of tinted acrylic sheeting can be fitted over the fibres; alternatively, two such domes may be used together to totally encapsulate the lamp, the lower one containing the light box and some form of stand. In either case the length of the fibres must be reduced, to bring the requisite dome radius down to manageable proportions. The method you use to mount the dome or domes will vary slightly with the size of the light box, but the smaller the light box, the easier the job will be;
—the ferrule that binds the bottom ends of the fibres together at the base must be firmly attached to the light box. Balance is the thing to watch for here: if the ferrule is slanted even slightly to one side the fibres will all tend to lean to that side, and if there is a sufficiently large bias will perhaps snap at their base. Remember that these fibres are of glass, and can pierce the skin if not treated with respect. In addition to this, they are far and away the most expensive single item in the lamp, so should be handled carefully at all stages of the assembly.

Components and construction

The light box for this particular design is made from three spun aluminium cake tins, which are freely and cheaply available in most hardware shops. Two should be of 7in. (175mm) diameter, and the other of 8in. (200mm) diameter, though the size will obviously depend on your individual design. The important thing is the relative clearance between the smaller and larger, or covering, sections, and about ½in. (13mm) around the inner section, as is used here, is recommended. Aluminium has the advantage also of being very easy to work with, so the numerous holes to be drilled in the light box will not cause much extra work. On the other hand, a container such as a copper bowl or a worked metal cylinder may appeal to you as more suited to your own decor. Unfortunately, you may have difficulty in finding the exact shape and size to suit your design if you decide on something a little more aesthetically pleasing than the aluminium cake tins.

The tins are cut and drilled according to the design on page 41. You will also need a number (about 10 should suffice) of self tapping screws. (about ¾in. (19mm) Gauge 8—tne precise size is not vital), some ½in. (13mm) screws with nuts and washers (about 10 of each), and 3 rubber ball feet to set the bottom of the light box on. It is also a good idea to use a rubber grommet to lead the power supply cable out of the light box itself without any danger of chafing or cutting of the cable covering. A Jubilee clip, of a diameter appropriate to the diameter of the ferrule at the base of the glass fibres, is used to bind the ferrule to the copper pipe sections which serve to attach the fibre base to the light box (see below for details).

Spacers made of small sections of aluminium tube are included in this design at three points (see the plans on page 41). You will need about 1ft. or 305mm. of aluminium tube of an inside diameter suitable for the size of screw used. Two brackets, made from a length of ¾in. (19mm) copper tube, are attached as shown in the step-by-step series to the aluminium cake tin that makes the base of the light box. Since this particular type of cake tin has a small fold of aluminium around its top edge, two slots must be cut at either side of the top tin edge where the brackets are screwed on—self-tapping screws should be used here. The brackets themselves are formed by splitting the splayed out copper pipe piece in two—leave about ½in. (13mm) of the pipe intact at the top to act as a clasp around the ferruled base of the fibres. The Jubilee clip is fixed around the bracket ends and the ferrule to provide a snug and secure anchoring for the mare's tail (see Fig.1).

The electrical equipment included in this design, and which in the main will be necessary no matter what your final design is, is as follows:
—one mains/12 volt AC transformer, which is connected to the light source
—one 5 watt, 12 volt (AC) bulb and its holder, of the type commonly used for automobile side and parking lights
—one 240 volt electric motor, of a type that will allow convenient mounting inside the light box with the drive shaft extending through the top of the inside section of the light box: one revolution per minute is possibly the best speed for capturing the full effect of the colour transitions
—sufficient three-core flex, fitted with a standard plug, to connect the lamp with the mains supply
—a short length of standard sheathed wire, for connecting the bulb and the motor to their respective power sources
—to assist you in making the flex connections inside the light box as tidy (and as safe) as possible, a small plastic junction block like that shown in the step-by-step series is a most worthwhile but inexpensive addition to this list.

The electrical components are arranged inside the light box as shown in the plans and the step-by-step series. Notice especially that the transformer casing, and the light box itself, is connected to the Earth core on the incoming mains supply cable. Spacers are used with both the motor and the bulb holder, in the first case to give the drive shaft unobstructed entry at the right height through the upper plate of the inner section of the light box, and in the second case to keep the bulb away from the bottom of the fibres—too close, and the ends will brown with the heat given off by the bulb. Once the light box has been assembled, you should let it run for half an hour or so and check on the amount of heat absorbed by the fibre ends to make sure that the distance between the bulb and the ferrule is sufficient to allow the ventilation produced by the holes drilled in the light box to minimize the danger of browning.

The colour wheel, which may need to be cut down a little to fit inside the light box (normal glass-cutters will be quite adequate for this), can be attached to the top of the drive shaft in a variety of ways, but since there is virtually no torque applied to it simply sticking it to the cog wheel on the drive shaft with a standard contact adhesive will be sufficient. You must obviously ensure that the colour wheel extends completely over the light source, as otherwise the fibres will be only partially coloured—in fact, one side of the lamp will be continuously glowing with 'white' light.

Further uses for glass fibres

As well as the mare's tail of fibres, which contains several thousand individual fibres, cables with around one hundred fibres are available, at proportionately lower cost. This type of cable is usually sheathed in a flexible plastic, which means that you can run a length of the cable from a light source and only expose the fibres for a couple of inches or so at the end in a miniature mare's tail. To decorate a Christmas tree, for example, several lengths of the thinner types of fibre cables can be run from one simple (or coloured) light source at the base of the tree to terminate in small mare's tails around the branches.

As well as their more specialized technical applications, these fibre cables have a variety of decorative uses—the most obvious one being a modified and simpler form of the mare's tail lamp described above, this time using a number of the thinner cables run from the one light source/base and finishing in smaller mare's tails of staggered heights. This could then be covered in a tinted acrylic dome, the dome itself possibly just resting on the table surface which also supports the lamp base. Another decorative application you might wish to consider uses the thinner cables to make small anemones of light at the bottom of a large fish tank, which will move with the currents in the water. In this case the light source, which again need not be coloured, or at least not multi-coloured, can serve all the individual cables. The light is positioned outside the tank, of course, and the cables run into the tank down the corner joins. You may have to experiment with the best length of fibre to expose at the end of the cable for your particular tank conditions. If you have quite a strong current in the tank, for example, too little of the fibres exposed and free to undulate in tune to the movement of the water may mean that the mare's tail will not be able to be seen to best effect.

Judging from the increasing popularity that these fibre optic applications are enjoying at present, further developments in both the scope and sophistication of the decorative items on this principle can be expected. The ideas suggested in this chapter are therefore just starting points—by keeping your eyes open to the do-it-yourself possibilities in this regard you may be able to hit upon an imaginative idea that will give you something a little different from the common run of interior decoration.

CAMERA PRESS

Something to glow about

An illuminated ceiling creates soft shadowless lighting and also acts as an insulating barrier, reducing the problem of condensation.

In addition, the panelling will conceal any ugly pipes or wiring round the walls. The panels are easily removed for ease of cleaning and maintenance.

The structure of an illuminated ceiling consists of a latticed framework, usually of aluminium angle strip, which is suspended a set distance below the actual ceiling. The framework is inset with translucent pvc panels which diffuse the artificial light source and create virtually shadowless lighting. It is a project that any competent handyman can install providing a few simple rules are followed and the correct materials used.

Planning

There are several points to consider when planning an illuminated ceiling. The space between the actual ceiling and the suspended framework must be at least 6in. or 150mm to allow both for the fitting of lighting units and for

ILLUMINATED CEILINGS LTD.

adequate dispersal of the transmitted light. So you must allow for a drop of at least 6in. in the height of your room. This may not matter, or may even be an advantage in the case of a high ceiling, but if your headroom is already low you will have to decide whether you can afford to lose the space. In extreme cases you might not be allowed to; in Britain, for example, the minimum allowable height for a room is 7ft 6in. or 2.15m and you will come out with less than this if your room is already less than 8ft high.

You must also assess your illumination requirements. The translucent panels will absorb approximately 50% of the light transmitted, so you will have to at least double the original output to achieve the same level of lighting. If you are dissatisfied with your present level of illumination, there is a rough rule of thumb for obtaining a suitable level using fluorescent tubes behind $\frac{1}{16}$in. or 2mm thick translucent pvc panels. An allowance of $1\frac{1}{2}$ watts per square foot or 300mm square will produce lighting sufficient for most purposes. As mentioned earlier, the overall diffusing effect depends on the distance between ceiling and frame—6in. is the minimum, but a greater distance than this will produce even better diffusing qualities.

It is usual to install fluorescent lighting behind the panels. Tungsten lighting can be used, but this source shows up as bright areas of light on the panelling under the bulb, creating a 'spotty' effect. If you want to install, or keep, a variable lighting intensity unit, bear in mind that the dimmer switches sold for this purpose are for tungsten lighting, and will not work with fluorescent tubes unless a choke component and cathode heating transformer are fitted.

The surface of the internal ceiling also has an important bearing on the amount of light transmitted. A coat of matt white emulsion is perfectly adequate, but high gloss white paint, or a covering of aluminium foil will provide a much more reflective surface.

Tools and materials

The framework for the panels consists of $1\frac{1}{2}$in. x $\frac{3}{4}$in. or 38mm x 19mm aluminium angle strip nailed or screwed round the wall; main support tees, $1\frac{1}{2}$in. x $1\frac{1}{2}$in., are laid across this, spanning the width of the room, and short $1\frac{1}{2}$in. x $\frac{3}{8}$in. bridging tees are placed between these. Fig.8 illustrates the three types of angle strip used.

Aluminium is used for this purpose because it is only a fraction of the weight of angle iron, yet sufficiently strong. The aluminium is sold in its natural finish or in a variety of anodized colours. If none of the colours is suitable, you can prime the metal with aluminium primer and paint it. Painting aluminium is not usually successful if the metal will receive a lot of wear, but this is not the case with ceilings.

Opaque pvc panelling $\frac{1}{16}$in. thick is available in a variety of colours and textured surfaces. This thickness is quite adequate, and is easily cut with a trimming knife—thicker panels will only absorb more light and provide unnecessary weight. As with the main support and bridging tees, the panels are just dropped in place and the whole ceiling is supported round the edges by the wall angle.

A bold modern dining room with soft shadowless lighting which creates subtle highlights on the chromed furniture.

GRAHAM TELLING

An illuminated ceiling fitted in a kitchen. The decor is reminiscent of a style that was popular in the late 1920s, and which is coming back into favour.

Measuring

The total length of wall angle required is equal to the distance round the walls, taking alcove recesses into account.

The total length of tee angle—both main support and bridging—depends on the size of the pvc panels; the larger the panels, the less tee angle required. Each main support should be equal to the width of the room and if, for example, the panels are 2ft square, you will require as many supports as are necessary to span the width at 2ft intervals. The bridging tees are worked out in the same way, except that these span the length of the room and are cut up into lengths to bridge either a pair of main supports, or a main support and a wall angle.

To estimate the number of pvc panels required, draw an outline of the room on a sheet of graph paper, using each square of the graph, or groups of four squares, as 'modules' of pvc panels. It will then be easy to count the number of panels required.

Fitting the frame

Mark a point on a wall at whatever height you have decided to place the ceiling. Hold a 3ft or 1m spirit level against the mark, adjusting the level until it is absolutely horizontal, and draw a line across the top. Repeat at 3ft intervals round the room.

Nail (with masonry nails) or plug and screw the 'L' or wall angle at 12in. or 300mm intervals along this line (see Fig.2). Where the angle strip is jointed at right angles, either going into a corner, or around one as on the outside of a chimney breast, mitre the angle with a hacksaw.

Preparing the ceiling

If you decide to line your ceiling with aluminium foil, there are several types from which to choose. Select the one that is most highly reflective. When buying the adhesive, ensure that it is the type that has an anti-fungicidal ingredient incorporated. The conditions inside a suspended ceiling—gentle warmth and relatively still air—are ideal for fungal attack.

The light source

At this stage the light source must be installed or laid out. If you have a single tungsten light in a small room, it is a simple matter to install a fluorescent tube fitting, using the existing wiring. But you are more likely to need two tubes to obtain better diffusion and adequate light. Two light sources can be wired on the same circuit, which means that they will both be controlled simultaneously by a single switch.

To install two lights which will operate from one switch, another rose is fixed to the ceiling and the wiring is extended from the old rose to the new one by loading a length of wiring from live to live, neutral to neutral and earth to earth.

Although the wiring is slightly more complicated, it is usually better to be able to control the two lights independently from a single point by installing a twin switch panel. To do this you will have to fix a junction box to the ceiling. The live wire is then taken from the junction box to the twin switch panel, then to the two ceiling roses. Earth and neutral are then taken from the box to both roses, and the earth to the switch, to complete the circuit.

Laying the ceiling

It may sound strange to talk about 'laying' a ceiling, yet that is exactly what you have to do to install an illuminated ceiling. Having secured the frame of wall angle round the room, all the rest of the structure is simply laid in place.

The main support tee sections spanning the width are dropped in to position, resting at each end on the wall angle. These main tees will span a distance of twelve feet with perfect safety in this fashion. If the width of the room is greater than this, special joining clips are used to join the lengths. In this case it is necessary to support the angle at the join by means of a wire secured to a hook fixed in the ceiling.

The bridging tees are placed between the support tees, creating a lattice of square tee strips except round the perimeter, where it is virtually certain that the panels will have to be trimmed to smaller, rectangular, sizes. It is highly unlikely that your room will consist of exactly square units all over.

The square pvc panels are simply eased in through the diagonal of each section of lattice, turned, and dropped into place.

Fig. 1
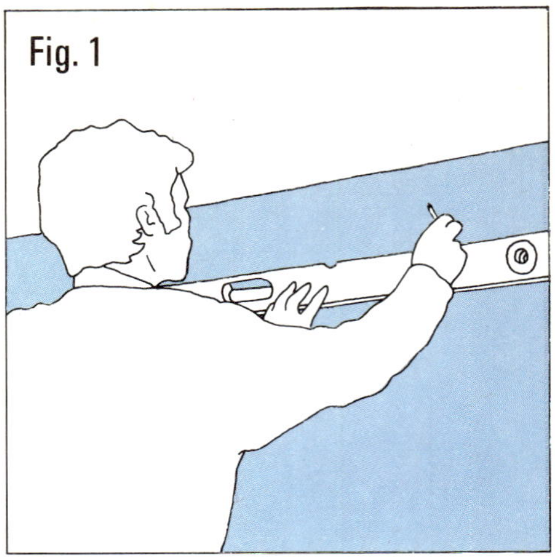

Fig. 2
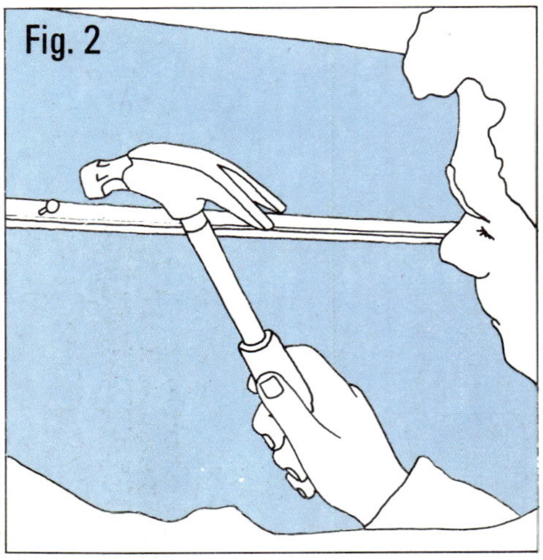

Fig. 3
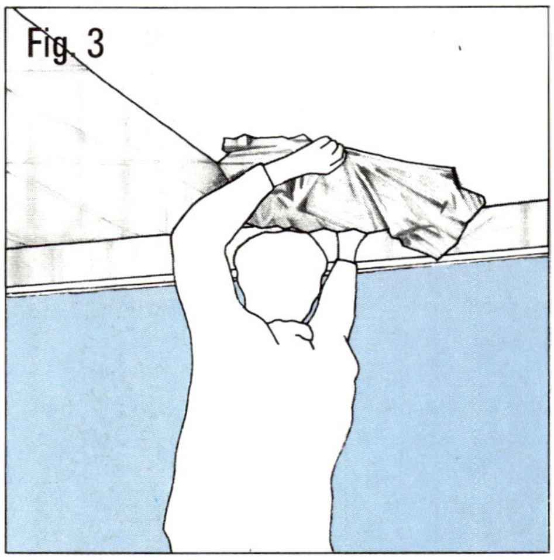

Fig. 4

Fig. 5

Fig. 6

Fig. 7
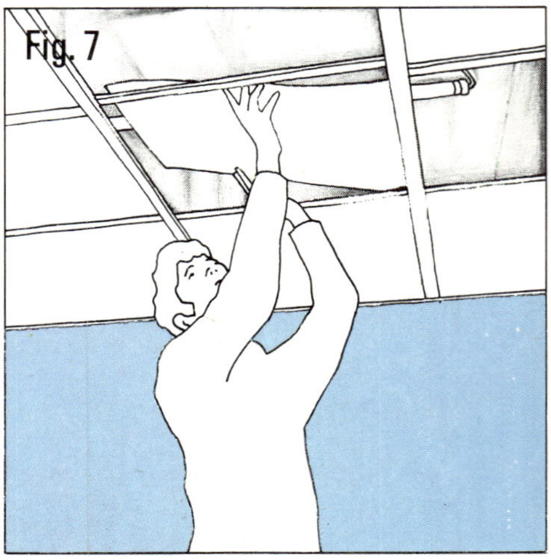

Fig. 8
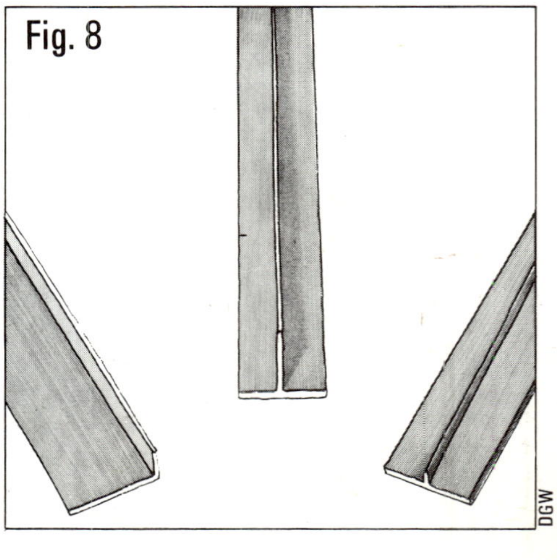

Fig.1. *Using a 1yd or 1m spirit level, draw a horizontal line round the wall at the required ceiling height.*
Fig.2. *Use masonry nails, or screws, to secure the wall angle round the room. This is the only section that is actually fixed.*
Fig.3. *For the best results, the interior should be lined with a highly reflective aluminium foil. If this is not possible, a high gloss white paint will do.*
Fig.4. *Fluorescent lighting is better than tungsten lighting, which shows up as bright patches of illumination.*
Fig.5. *The main support tee angle is simply laid across the narrow width of the room between the wall angle.*
Fig.6. *Placing the bridging tees in position to complete the latticework, ready to take the panelling.*
Fig.7. *The translucent pvc panels are simply eased through the diagonal of each square and dropped in place.*
Fig.8. *The three types of angle that make up the latticework of the ceiling. Left to right, wall angle, main support tee, and bridging tee.*

The magic that's all done with mirrors

Without a looking glass a room is only half alive. From great sheets of glass that cover a whole wall to the tiny little Victorian nonsenses, framed in shells, that you sometimes come across in junk shops, they all have a part to play in adding a touch of glamour to a room.

Mirrors are space-makers that are able to give the illusion of added dimension; decorators and reflectors of light; objects of light and beauty when properly used—but capable of appearing as ugly holes in the wall when their function is misunderstood or abused.

Mirrors as space-makers

Mirrors create space through illusion, by deluding you into believing that where you know there is a wall there is space. Properly used, these slabs of mirror make you feel that you could walk straight through them into an unexplored dimension.

To do this successfully the mirror has to be placed where, structurally, if there were no wall, you *could* walk on. It is no good putting one beside a window which looks out over the garden. The mirror will reflect the room and, since the eye is no fool, it will do a double-take when it sees garden and room in the same dimension. For a moment you will be surprised, then you will realise the deception. But, placed on an inner wall where you could expect more rooms beyond, a mirror will have a quite different effect.

Below. *These double doors covered with mirror glass create an illusion of extra light and space in this room.*

HEIDEDE CARSTENSEN

These deceivers should be used so that there is a reason for their end and their beginning. For maximum impact, they should begin at the floor (not at the skirting, because the floor should appear to run right into them), reach up to the ceiling and to the end of the wall on both sides. If you do not want a mirror that wide, or if it does not reach from wall to wall, you must provide a reason for its not doing so.

One way is to set the mirror into panels that cover the rest of the wall. These panels can either be made of wood or be papered to match the rest of the room. Another way is to curtain the rest of the wall, hanging the curtains on rings suspended on brass poles fixed at ceiling height. Both of these effects make the mirror appear to be an integral part of the wall, not an afterthought added at random. A large framed mirror will also be infinitely more satisfactory than fixing a plain sheet of glass to the wall. The framed image is very much more telling than that which, for no apparent reason, suddenly stops with the edge of the glass. You are dealing with illusions, remember.

Line the alcoves on both sides of a fireplace with mirrors, from floor to ceiling and wall to wall, then—providing that you arrange your possessions so as to give a satisfying reflection—you will feel that you can walk straight past the chimney breast into the space beyond. Or, if your living room seems to come to an abrupt end in a hard-looking wall, try putting a slab of glass in the middle of the offending wall. Set it in panels, or curtain up to its edges, possibly adding a swag or two of the material over the top, rearrange the furniture until the reflection pleases you, and your blank wall will disappear. This is a good treatment for a dead-end wall at the end of a hall.

But if you do feel like covering one whole wall of your living room with mirror glass—and this can be most effective—it is really better that it should be the wall *behind* the sitting area rather than the one *facing* it. It can be very distracting to have to watch your own reflected antics for an entire evening.

Experiments

The bathroom, or a tiny cloakroom without a window, are perfect arenas for more adventurous experiments with mirrors. Try lining all the walls of the bathroom with them (forget what was said earlier about window walls—the eye will be far too busy staggering about in infinity to worry about that) and mirror the ceiling as well. As long as there is something worth reflecting, you can transform a dull room into a fascinating kaleidoscope of images.

Do not be put off by the bathroom fitments. There is nothing intrinsically ugly about a

HEIDEDE CARSTENSEN/PHOTO STUDIO DIE WOHNFORM

lavatory or bidet. The shapes, as shapes, are quite pleasing aesthetically, but familiarity has bred contempt. An attractive floor covering of either carpet or tiles, decorative jars or plants and pretty towels will all add to the reflection.

Or, if you feel that this is overdoing it, a sheet of mirror over the vanitory unit will create light and space as well as look good. Fix it hard up against the walls and to the top of the unit, and then try another on the wall behind you.

Below left. *The use of mirror glass along one wall of this dark, narrow hall makes it look brighter and wider.*
Below. *Smoked mirror glass on sliding doors gives a subdued reflection in this study, and the smoked glass partition enclosing bookshelves adds to the illusion of space.*
Below right. *A circular skylight above a landing lightens the whole stairwell.*

Mirrors as reflectors of light

Mirrors can be invaluable in helping to lighten and brighten dark corners or even whole rooms. Two mirrors placed opposite each other, butting up to the window wall, for example, will give a room the illusion of greater width and will also make it lighter, by 'bouncing' the light further back into it. This is a very good way of brightening up a long, narrow bedroom, or any dark, confined area such as a hall.

If you panel the sides of a dormer window, in an attic, for instance, with mirror glass, the natural light will be greatly increased by reflection. Lining the whole deep reveal of a window with mirror glass will have a similar effect.

A little cubby-hole of a room, say a cloakroom, or any dark place with a hopelessly inadequate window, will be transformed if you cover the ceiling with mirror glass and let it reflect light from strip or spot lights.

Why use glass mirrors?

Glass, lovely as it is, is dangerous stuff to work with, as well as being very heavy. But you need not use glass at all. Silvered acrylic sheeting comes in sheets of up to 8ft x 4ft, the mirrored reflection is undistorted, it is half the weight of glass, and it is practically unbreakable. It can be cut with an ordinary saw and fixed with screws or with a special sticky tape. It comes in fifteen colours, including colourless and some stunning brownish colours. It is also almost immune to condensation. This seems the ideal material for a mirrored door that saves so much space in a tiny bedroom. There is even a specially toughened grade that is bullet-proof; hardly necessary in most homes, but it should be able to withstand the rough treatment reserved for nursery walls.

There is a semi-rigid pvc sheeting which can be stuck on flat surfaces to produce a fairly good

JULIAN SEDDON/DAVID ESSER LTD.

CONWAY PICTURE LIBRARY

BRIAN MORRIS

image. This might be used on ceilings, as it is light and easy to handle. It can also be stuck to curved surfaces, when it produces those fascinating distortions usually only to be found in fairgrounds. A mirrored material made of very strong terylene treated with a special reflecting product is also available. It is stretched on an aluminium frame over a resilient backing board. The result is a gleaming, perfect mirror surface that is unbreakable in the way that glass is breakable. It is, however, easily punctured or cut, therefore its use should be limited to situations where this is not likely to occur. This material is ideal for ceilings; its virtues for that purpose being its extreme lightness and the fact that it is immune to condensation. It is easily fixed by means of impact fabric pads (such as Velcro). It comes in 15 standard sizes, or specials up to 4ft wide can be made to order. It is impervious to heat from lights, and apertures to take ceiling lights can be cut into specials.

Mirrors used for decoration

Here is a list of suggestions for more sites for mirrors, used for decoration and sparkle, rather than for space-making.

1. An alcove behind shelves. This has been done before, but try using coloured, engraved or etched glass for a new effect.

2. To block out a window where the view is really impossible. Fill in between the glazing bars with mirror glass, and light with strip lights hidden behind the curtains or a pelmet.

3. At the end of a dining table that is pushed hard up against a wall. Again, etched or faintly coloured glass can look good.

4. Framed, as a bedhead. Be careful, however, to fix it properly. Silvered acrylic would be the safest.

Glass

Many people forget that windows are not absolutes. They can often be added or enlarged. Small windows in the garden wall of a ground floor living room, if converted into French windows, can open up a pleasant view as well as giving access to the garden. If you are a high-level flat-dweller with no garden, a new window opened into an outside wall may reveal a hitherto unseen roofscape.

If you live in a terraced house where window walls are limited, you may like to cut a skylight in the ceiling. If fitted with attractive roller blinds it can look most effective, and let in enormous amounts of light. Attic rooms often have only tiny windows. These can quite easily be converted into dormer windows, giving a room not only more light, but also more space.

It is amazing how much warmth the sun can give to a room, even in winter, if a window is well placed. Sun rooms, or just south-facing picture windows, let in the maximum amount of sun and create a feeling of extra space, particularly if they open onto a garden.

Interior uses of glass

Glass has many interior uses which are not always appreciated, and seldom exploited. You may want a glass shower screen to replace the plastic curtain that always looks so tatty. You should use, for safety, toughened glass about 5mm thick which, if smashed, shatters into small, relatively harmless fragments, as does a windscreen. This toughened glass comes in good patterns, which tend to disguise splash. It is wise to use this type of glass for glass doors, especially if you reflect that 80% of all the accidents with glass are caused by doors. Very beautiful doors, made entirely of this patterned glass, 10mm thick, can be bought. Some patterns should please the architecturally minded, and others, including a naturalistic pattern of leaves, are for the less austere. These doors, which are translucent but not transparent, would be beautiful for an inner front door, or, indeed, for any door leading off a hall.

Room dividers and screens

Coloured or patterned glass is also suitable for panels in room dividers, or for any fixed screen. If its purpose is to separate a kitchen from a dining area, coloured or heavily patterned glass helps to hide the ugly sights of the kitchen while you are eating. However, if the divider is only separating the dining area from the sitting area in a living room, then it does not need to be so opaque, and plain or pale-coloured glass is all that is needed.

For the really ambitious amateur glazier there is marvellous 'antique' type glass, such as is used in stained glass windows. It is thick, lovely and expensive, but can have many imaginative uses. There is even chunky, coloured glass, suitable to be set in concrete or epoxy resin. Decorative exterior screens of tough glass can act as a windbreak beside a patio, or hide dustbins or a garage wall.

Glass shelves

Many people have collections of glass ornaments; these can look out of place arranged on heavy wooden shelves, and are also difficult to light effectively. Glass shelves built in alcoves can look most attractive, and with strip lighting above can give an extra light source to a room. Delicate ornaments which collect dust easily can be protected by shelves which are 'glassed in'.

It is important to remember, whenever and however you are using additional glass in your home, that—decorative as it can be—it is a cold material. Wherever you are putting it, the room should be adequately heated. Enlarging the window area in a room may let in more light, but it will certainly let in more cold as well.

Look around your home with these ideas in mind, as you will be surprised how many rooms can be improved by the addition of a mirror or some glass. Your success lies in imaginative application.

Coloured, frosted or etched glass can be used effectively in bathrooms, either to hide an ugly view without obstructing the light, or just for decoration. ***Top left.*** *A rural scene painted on a bathroom window.* ***Centre left.*** *The frosted glass window is reflected in the etched glass mirror above the bath.* ***Below left.*** *Cut and etched mirror glass produce an ornate, Edwardian-looking bathroom.* ***Right.*** *A picture window overlooking the country lets in lots of sun and adds an extra dimension to the room.*

An illusion of space

It's never easy working within the confines of a small flat, yet there are many techniques which can create a feeling of openness and space. One of the most effective is the use of mirrors. The dimensions of any room seem to expand through cleverly placed mirrors. This ninth floor flat mirrors the world outside and makes you totally unaware of its actual limited size.

Hat maker George Malyard designed this flat —his twenty-first. His main objective was to get away from the pressures of the chaotic world. He succeeded in finding a flat in central London which was far enough away from noise and car fumes; it was nine floors above the congestion and commotion of the street.

The main consideration was how to eliminate the feeling of being cooped up in two little garret rooms. The answer to this problem was solved with ceiling to floor mirrors on the walls. The effect is miraculous—the illusion of space in the flat continues as far as the eye can travel looking out of the window. There seems to be no division between the world without and the world within. There is an added dimension.

Mirrors are invaluable if you want to 'open up' a room. Give careful consideration to where you put them. They should pick up the maximum amount of light available in the room, and also reflect an interesting view or objects. Before choosing the spot for your mirror, try sitting in different parts of the room and see where the mirror can have the greatest impact. Try placing the mirror on different levels and at various angles. If you're in the process of installing mirrors, the preceding chapter should be some help on the practical side.

As Mr. Malyard looked out the flat windows, the colours he saw were the greys and silvers of concrete buildings and the reds of tiled roof tops. His colour scheme was decided. He would bring the colours from the immediate outside world into his own flat. Grey, silver, terracotta and white are the dominant colours. The designer didn't want to live with a view, but *in* a view.

The entrance lobby leads into the sitting room, which measures approximately sixteen feet by twelve feet. Hues of terracotta emerge from the Italian leather button-backed chairs and the Spanish floor tiles. The unusual curves in the tiles echo the curve over the door. The settee is a dark brick colour, made from coating fabric with central insets of Persian rug. The Persian motif is repeated in the floor rug which has similar patterns and colour tones.

The remaining objects in the room do not have a fabric or leather finish, but a hard metallic one —just like the chrome and concrete in the world seen outside. The transparent tables are solid perspex; the dining chairs are perspex and steel. The sliding windows are framed by part-polished and part-lacquered aluminium. During the day, the drab, fog-coloured blinds—again an echo from the outside—can be used to give a

Below. *A small room with a big look. Floor to ceiling mirrors capture the panoramic view, making the room seem far more expansive and open. Beluchi rugs serve as couch insets.*

JOYCE WREFORD/RONALD A. CHAPMAN

***Above.** Looking out of the ninth floor window. Steel and chrome outside echo the aluminium window frames and perspex table and chairs. Another mirror serves as a drinks table top.*

restful green colour in the room. At night, blinds aren't needed. In fact, there is enough light from the city to sit and relax in the room without any indoor lighting. George Malyard finds looking at the panorama around him far more interesting than television.

The ornamentation is in keeping with the metallic look. The picture frames are solid white gold leaf. The silver sculpture on the window sill is a polished aluminium alloy. One modern Italian lamp has a silver finish and the other, an aubergine finish. The designer made the sculpture and also painted some of the wall pictures.

Mirrors are used again in the bedroom. It consists of a double bed set into an alcove, surrounded by mirrors, and two big walk-in wardrobes with mirrored sliding doors. The walls are covered with a mushroom brown imitation suede which is actually waterproof raincoating. The delicate print of the continental quilt cover contrasts well with the more subdued brown-champagne colours. The trim around the alcove and the wardrobes is nail-studded chrome and steel. There is a surprise colour—a butterfly blue—carpeting the wardrobe floor. The teak shelves in the wardrobe are set against printed silver foil in terracotta and grey.

The compact kitchen repeats the metallic and white motif. It is well fitted with a built-in eye-level oven and grill and a waste disposal unit. The refrigerator neatly fits under the oven. The cupboard surfaces are covered with dark blue melamine.

Silver-grey colours also dominate in the bathroom in the tiles and vinyl floor. The design has been kept simple to eliminate the cluttered look.

The flat has been designed for labour saving. All surfaces can be cleaned with a cloth except for the Persian rugs and the suede-like bedroom wall covering, which can be vacuumed with a small cleaner. Leather cream is used on the terracotta chairs, and the perspex tables are rubbed down with anti-static polish.

Another feature worth noting is the lighting. Most of the lights are spots and have dimmers to save on the electricity bills. Dimmers are practical and can have a dramatic effect. They can be adjusted according to the time of day and your mood. Thev are useful in a room where study is done some evenings and television watched on others. Lighting for living rooms, gives lots of ideas for various choices of lighting available.

From one small flat, George Malyard has created a spacious looking home through the miracle of mirror and design. If you are confined with rooms that make you feel closed in, remember that clever placement of mirrors can make all the difference to enchancing the room as well as making it look larger.

***Below.** A striking silver sculpture and two ultra modern lamps overlooking the city set a streamlined urbane tone to the room. Mirrors again reflect the night lights outside.*

JOYCE WREFORD/RONALD A. CHAPMAN

Illusions of grandeur

The use of mirrors is an unbeatable way of creating the illusion of space in a small room. It is not essential to use large glass mirrors; in the right places, mirror tiles and silvered acrylic sheeting will do as well or better. Here is how to fix all three kinds of mirror.

There are two main points to consider when choosing which type of mirror to buy: the price and the quality of the reflection. Obviously, you want to pay as little as possible for your mirror, but the unsatisfactory reflection given by the cheaper kinds limits their use to places where the hazy image does not matter. For example, a mirror used on the end wall of a dark passage to lighten it need not give a clear image, because its purpose is only to reflect light that would otherwise be lost. But a mirror that is a central feature in a living room has to be made out of high-quality glass, because people will constantly be looking at their reflections in it.

Sheet glass mirrors

Ordinary mirrors made of float glass give the highest-quality image of any kind of mirror, and are the normal type used in most situations. They have two disadvantages, however. For one thing, they are expensive, so that covering a whole wall with glass mirror is out of most

Below. *The mirror tiles on this kitchen wall give a whole new dimension to the room with their fascinating broken-up reflection.*

TRANSWORLD

people's financial range. For another, they are fragile, so that if you mount one on, say, a wardrobe door, you are taking a risk of breaking it every time you shut the door. This risk can, however, be lessened by mounting the mirror as firmly as possible on a strong, solid door.

The traditional way of mounting a mirror on a wall is to frame it and hang mirror and frame up like a picture. You can make the frame yourself or use an old picture frame and cut the mirror to fit the frame. But if you are mounting a mirror to make a small room look spacious, it should not be put up in this way for two reasons. One is that the frame makes a clear line of demarcation between the real room and its reflection, reducing the effectiveness of the illusion. The other is that a framed mirror hung on a nail with picture wire hangs at a slight angle, and the reflection is tilted twice as much. This spoils the illusion of space completely.

The solution is to use a sheet of unframed mirror hung on small, neat proprietary clips that hold it firmly against the wall. Any glazier will cut you a piece of mirror to the exact size you require, without charging much for the job.

Most walls in modern houses will be flat enough to mount a large mirror on directly. In a house with noticeably curved or lumpy walls, you should mount the mirror on a sheet of $\frac{3}{4}$in. (19mm) blockboard exactly the same size as the mirror. This sheet should be painted or varnished on both sides and the edges to stop moisture from warping it.

There are three main types of support for mounting glass mirrors, each of them suitable for different mirror sizes and situations. The first of these is the *mirror screw,* which passes through a hole in the mirror into a plug in the wall behind (see Fig.1). Your glazier may agree to drill screw holes in the glass for you. If not, you can do it yourself with a spear-point drill, following the instructions in the following chapter. Use water as a lubricant instead of turpentine, which may spoil the silvering.

Under their removeable chromed heads, mirror screws have conventional countersink heads. To avoid the need to countersink the drill holes, drill them larger than the shank of the screw (but not, of course, larger than the head). The screw head will sink part of the way into the oversized hole, and the chromed head will disguise the rest of the projection.

Mirror screws are suitable for small or medium-sized mirrors up to about 4ft x 3ft (1.2m x 0.9m). Any mirror larger than this will put too much strain on the screw-holes and may crack the glass. Screws should be inserted at each corner and at approximately 1ft (300mm) intervals along all four sides, not less than 1in. (25mm) from the edge of the glass. When you have drilled all the screw holes, hold the mirror up to the wall and mark through the holes with a pencil to find the place to insert the wall plugs.

Mirror corners are much less likely to crack the glass, and can be used for larger mirrors (but see below). They are simple chromium-plated corner supports fitted with two eyelets that allow them to be screwed to a wall or other flat surface behind the mirror (see Fig.2). Corners of this type are very simple to mount.

TIOFOTO

Above. *The unusual 'hammered' effect of this mirror wall could be recreated quite cheaply by using silvered pvc sheeting.*

The mirror is held against the wall in the position it will occupy, a mirror corner is placed over each corner of the mirror in turn, and the wall marked for the insertion of wall plugs by sticking a pencil through the two eyelets.

Corners, which are silvered to match the mirror, are reasonably unobtrusive, but it is impossible to hide the heads of the screws that hold them on. Furthermore, they cannot be used to mount a mirror on a blockboard backing, because the eyelets are parallel with the face of the mirror. Both these problems, however, can be solved with special small *mirror clips* of the types known in Britain as 'Verity' and 'Valba' (Fig.3) and *mirror edge clips* (Fig.4).

'Verity' clips have two forms: one for the top of a mirror and one for the bottom. In both types, the mounting screw is concealed behind the mirror, and the clip is hooked around the edge of the mirror. The bottom clip has a conventional round screw hole, so that it is held rigidly to the wall and does not slip. The top clip, which carries no weight, has a long slot so that it can be slid up and down.

'Valba' clips are much the same, but have clear plastic facings instead of chromed metal ones. These may be less visible in certain locations. The sliding top clips are spring-loaded, which makes them less likely to come undone.

To mount a mirror on either type of clip, hold it up to the wall and draw right round it lightly with a pencil. Take the mirror down. Plug and screw the non-sliding bottom clips to the wall at 1ft (300mm) intervals so that the part that fits around the mirror is level with the pencil line. Then fasten on the top clips so that when they are slid up as far as they will go, the hook at the front of the clip just clears the pencil line. Mirrors over 2ft (600mm) tall should have 'top' clips down the sides as well, at approximately 2ft (600mm) intervals.

When all the clips are in place, lift up the

HEIDEDE CARSTENSEN/STUDIO DIE WOHNFORM

Above. *Conventional mirror tiles lighten this kitchen alcove, and are more decorative than a conventional float glass mirror.*

mirror and rest it on the bottom clips. Then press the mirror against the wall and hook the top and side clips around it by sliding them down and inwards. If you are using spring-loaded clips, you will need a helper to hold them back while you lay the mirror flat against the wall. In any case, make sure all the clips are hooked firmly in place before you let go of the mirror.

Mirror edge clips are used for holding a mirror to a board backing the same size as the mirror. They are fixed by laying the mirror down on the board, placing the clips in position one at a time, and screwing them to the edge of the board. They should be placed at 1ft (300mm) intervals along the top and bottom, and twice as far apart up the sides. But for very large mirrors over 6ft high, see below.

Corners and clips of all types are suitable for quite large mirrors, say up to 5ft x 4ft (1.5m x 1.2m). Above this size, however, the sheer weight of the glass may cause the clips to bend, or the corners to break the glass. With mirrors mounted on a wall, the problem can be solved by using corners *and* clips, and spacing the clips along the bottom at 6in. intervals. If this sounds messy-looking, you could substitute a piece of standard aluminium channel strip for the bottom row of clips. This should run the whole width of the mirror. to provide the best possible support.

The strip should be of J-section, as shown in Fig.5, to make it possible to screw it invisibly to the wall. If you cannot find any strip of this shape at your local metal merchant, buy a piece with sides of equal length and improvise a cutter, as shown in Fig.6, to slice part of the front off. This cutter should be made out of a wood block with a deep channel cut in it to fit exactly over the aluminium strip laid on its side. Cut a slot in the wood along the floor of this channel and glue a small 'junior' hacksaw blade into it. The blade should be more than half buried in its slot for rigidity, and exactly parallel to the sides of the channel. Use the strongest possible glue, preferably epoxy resin. And when you use the cutter, bang it to get rid of metal particles after every few strokes. This will help avoid any tendency to 'wander'.

This improvised cutter is the only tool that will cut channel strip in this way. An ordinary hacksaw will not, because the frame gets in the way. A special sheet saw is too large and innaccurate.

If you are mounting a very large mirror on a board or door, you can save yourself all this trouble by using L-section aluminium strip hooked round the front of the mirror and screwed to the edge of the board. Note, though, that this will not work where the screws are going into end grain, nor on low grades of particle board.

Mirror tiles

Mirror tiles are much cheaper than one-piece glass mirrors, and easier to apply. If you do not mind the reflection in them being broken up into squares, they are a very good solution

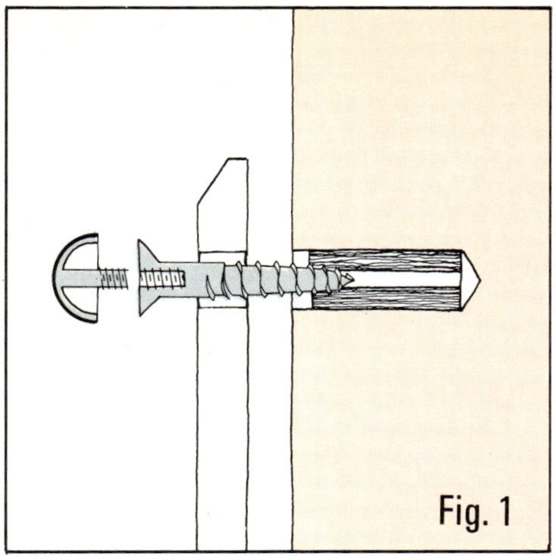

Fig. 1

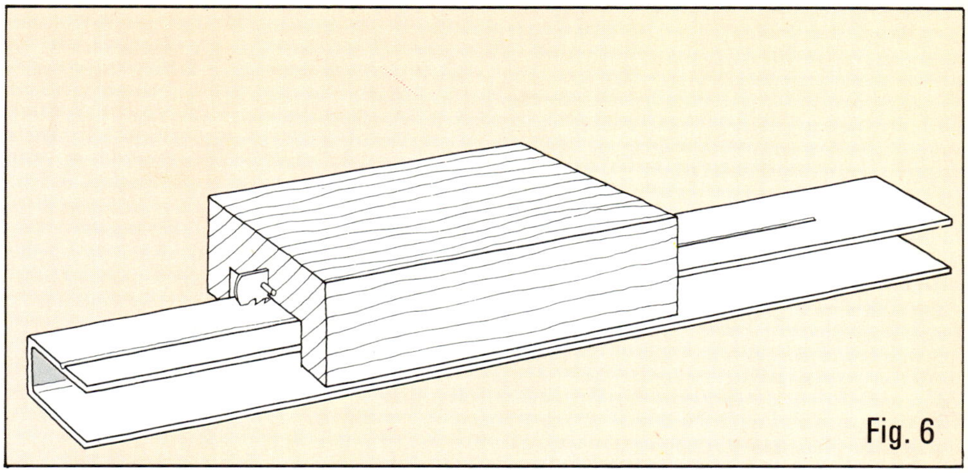

Fig. 6

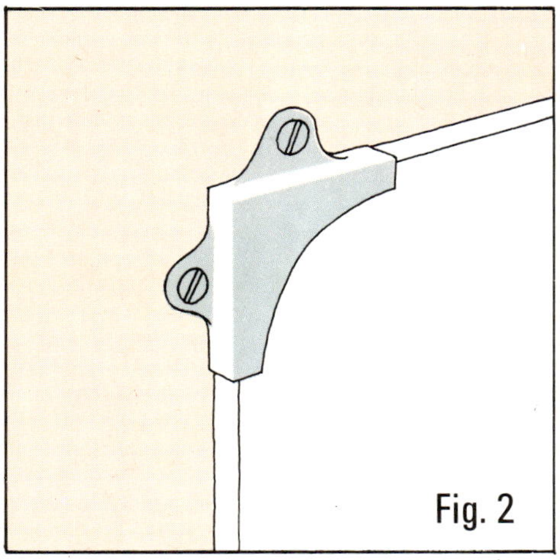

Fig. 2

TRANSWORLD

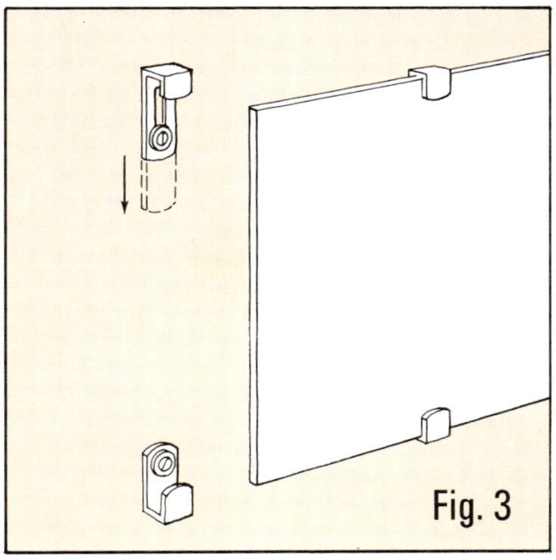

Fig. 3

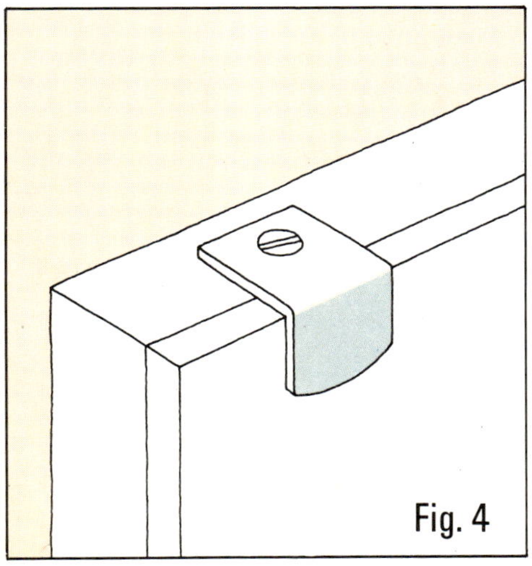

Fig. 4

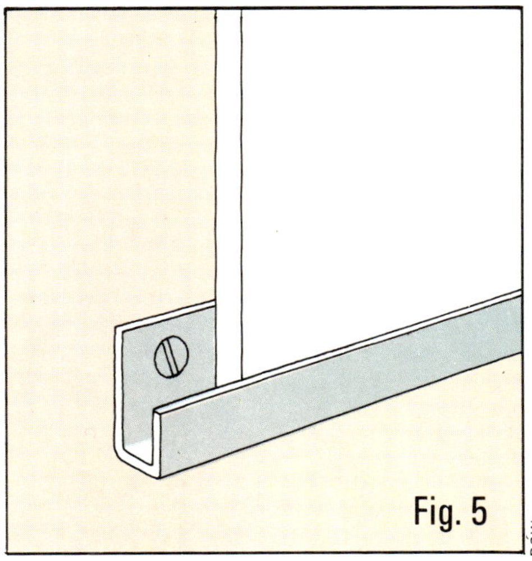

Fig. 5

DGW

Large picture, above.* *The reflection seen in the mirror wall on page 54. If this is too 'jazzy' for your taste, use larger tiles.

Fig.1. *(top left). A cross-section through a mirror screw, showing how the domed head screws into a hole in the main screw.*

Fig.2. *A mirror corner, which can be used to support larger mirrors than a screw.*

Fig.3. *Mirror clips will support any size of mirror, if you use enough of them.*

Fig.4. *Mirror edge clips look the same from the front, but fix to the edge of a board.*

Fig.5. *Very large mirrors can be supported by J-section aluminium strip as shown here.*

Fig.6 *(top right). Ordinary U-section strip can be cut to the right shape with a tool made from a hacksaw blade set in a wooden block. The lugs on the blade stop it from slipping.*

to the problem of mounting a mirror on a wall. Many people even find the broken reflection interesting, like a cubist painting.

Mirror tiles are mounted on little foam plastic pads with a pressure-sensitive adhesive on both sides. This makes the mounting slightly flexible, which is a useful safety factor. The standard-sized small tiles are attached by one pad at each corner. Some manufacturers make larger, rectangular tiles that need six or even eight pads to hold them on, but all types come in packets with the manufacturers' instructions, so there should be no confusion.

Calculate the number of tiles you will need exactly, as you would for ceramic tiles. Your stockist will advise. The tiles are cut exactly like window glass, using an ordinary glass cutter. The little pads will only stick firmly to a glossy surface, so emulsion-painted walls should be given a coat of gloss all over the area where the tiles will go.

Each tile should be applied by pulling the protective paper off *one* side of four self-adhesive pads (or however many you need). Do not touch the sticky surface with your fingers. Press one pad into each corner, then remove the paper from the other side and press the tile on to the wall firmly, but not too hard, or you may break the glass.

When applying tiles to irregular walls, one side can be brought out a short way by sticking two pads one on top of the other. Very irregular walls may require the use of adhesive as well as the pads, but use *only* an adhesive recommended by the manufacturer. British examples of suitable adhesive are 'Mirra-Mastic' and 'SG2'. Ordinary contact adhesive or tile cement may damage the silvering.

Silvered plastic sheeting

There are two kinds of silvered plastic sheeting. Silvered acrylic sheeting is slightly, but not very, flexible and gives quite a good reflection. It comes in many colours and textures, and two thicknesses: $\frac{1}{8}$in. (3mm), used for most purposes, and $\frac{1}{4}$in. (6mm) which is less likely to bend and gives a reflection of a very high standard. Both are virtually unbreakable, but scratch easily. They are very expensive; even the thinner grade costs considerably more than ordinary sheet glass mirror.

Silvered pvc sheeting is much cheaper, but gives a hazy and distorted reflection. It is very flexible and can be cut with scissors. It is suitable for mounting around columns, or for forming small silvered panels in a larger design, but it is useless for covering large areas unless you actually *want* a blurred, wrinkled reflection.

You may wonder why anybody bothers to use either type of silvered plastic sheeting. But both do have advantages. The heavier type, when correctly installed, makes quite a good mirror with two important advantages. It is unbreakable, so it is entirely safe to use on wardrobe doors (its lightness is also an advantage here). And it conducts heat much less efficiently than glass, so it is less prone to condensation in the steamy atmosphere of bathrooms—an important point this, as anyone with a glass bathroom mirror will realize.

The thinner pvc sheet is even lighter, very simple to cut out and install, and can be curved around anything. Provided you realize that it is no substitute for conventional mirrors, it is a versatile decorative material.

Acrylic sheeting can be cut to size by the shop, who will charge you a small sum for the service. If you want to cut it yourself, you can use a handyman's knife and steel ruler for the thinner (3mm) grade, and a fretsaw, hacksaw, sheet saw or fine-toothed tenon saw for either grade. Do not remove the protective paper when cutting—in fact, leave it on until the very last moment, when the mirror is actually installed.

Mount acrylic mirror with mirror screws (but not clips) in the conventional way. The holes should be drilled with a blunt twist drill, drilling slowly and not pressing very hard, so as not to melt or crack the sheet. If you want to keep the sheet absolutely flat for the best possible reflection, make it a backing of $\frac{1}{2}$in. (13mm) chipboard. The mirror should be fastened to this with double-sided tape or special acrylic mirror adhesive ('Mirofix' is a British example). Do not use any other kind of adhesive, and do not try to peel either tape or adhesive off once they are in place—you will almost certainly take off the silvering.

Silvered pvc sheeting should be attached with tape or 'Mirofix' in exactly the same way. There is no point in fastening it with screws.

HEIDEDE CARSTENSEN/STUDIO DIE WOHNFORM

Above. *Ordinary glass mirror is particularly effective used behind display shelves in an alcove. Here, glass shelves allow even more of the exhibits to be seen.*

Let the sunshine in

Glass is one of the most versatile building materials. Modern glass production offers a wide range of glass for glazing purposes, both plain and decorative, which are relatively inexpensive, and easy to obtain.

Simple glazing techniques requiring the use of few tools are quite simple to learn, and when coupled with a basic knowledge of the different types of glass and their uses, will enable you to

PICTURES COURTESY OF PILKINGTON BROTHERS LTD.

use glass to its best advantage in decorating.

Special considerations

All building materials are to some extent controlled by building regulations and local by-laws. In new building works and where any alteration is carried out, the size and the type of glass and its position in the building may have to be considered with regard to fire resistance, heat and sound transmission. If in doubt, consult the local building authority.

Choosing the glass

The thickness of the type of glass required is always related to the size of the sheet, and to the degree of exposure to wind and consequent suction loads on the surface. Manufacturers produce readily available tables which give the correct weight and thickness of glass to be used for particular purposes. Most glass used for domestic glazing in Britain is 4mm (about $\frac{1}{8}$in.) thick.

The glazing quality of the glass you are using is also an important consideration to make. Sheet glass has three grades: OQ-Ordinary Quality; SQ-Selected Quality; SSQ-Special Selected Quality. For general glazing purposes, the OQ grade is suitable.

Types of glass

Sheet glass

Used for normal glazing work, this is a clear, drawn glass. Since the opposite sides of a pane are never perfectly flat, and parallel, some degree of distortion is inevitable.

Sheet glass is commonly used for domestic glazing and is available in thicknesses from 3mm (just under $\frac{1}{8}$in.) to 6mm ($\frac{1}{4}$in).

Float glass

This type of glass is made by floating molten glass on a bed of molten tin. Such a process produces a glass similar to plate glass, but its higher quality dispenses with the need for additional polishing and so it has generally superseded plate glass.

Float glass comes in thicknesses from 5mm ($\frac{3}{16}$in.) to 25mm (1in.). As this glass is quite free from distortion and is strong, it is often used for domestic purposes, especially for large picture windows and table tops.

Float glass is also produced in plain or tinted decorative forms, by using various processes including acid etching, electrofloat and sandblasting. Additional decorative effects are obtained by the introduction of certain materials into the molten glass mix to alter its light and heat transmission qualities.

Mirrors are also made from float glass of the SQ grade, which has been moisture-proofed, silvered and edged in a variety of ways by machine and hand grinding.

Patterned glass

This range includes glasses with several different decorative finishes which are frequently used for 'modesty' purposes in bathrooms and toilets. They are specially designed to allow light through while obscuring vision. One typical example is rough-cast glass, which is smooth on one side and obscured on the other.

These glasses, which are usually produced by passing liquid molten glass through rollers, are available in a variety of styles such as *artic, cotswold, reeded, autumn, flemish* and *patchwork*, in either plain or tinted versions. The degree of transparency and light diffusion desired will largely determine the type to be used. Thicknesses generally are between 3mm. ($\frac{1}{8}$in.) and 5mm ($\frac{3}{16}$in.).

Wired glass

This glass has a metal mesh embedded into it to reduce the risk of injury from falling bits of glass. It is also accepted as a fire retarding material. The mesh may be of a square (Georgian) or diamond pattern.

This type of glass is often used in porches and garage roofs where breakages may occur, because the wire mesh helps to hold the glass together if it is broken.

If you intend to re-use this glass, you may find it difficult to cut, and for this reason it is advisable to buy it already cut to size from a merchant. Wired glass is available in 6mm ($\frac{1}{4}$in.) thickness only.

Toughened glass

Float, sheet and even some patterned glasses can be toughened by a special process which can increase the strength of the glass by four or five times. Is particularly suitable to use for doors and similar areas where there is a danger of impact, since it shatters into granules instead of sharp splinters if it is broken.

Profilit glass is a strong, translucent glass used mainly outdoors for carports and the like, although it can be used indoors.

Solar control glass reduces the transmission of heat, light and glare from sunlight. It comes in a range of colours in float, laminated, rough-cast and patterned form.

1

Stack glass at a shallow angle against a wall with protection at the sides and bottom.

2

The proper method for one person to carry a sheet of glass—note the protective gloves.

3

To cut a straight piece, check to see that the glass edge is true with a T-square.

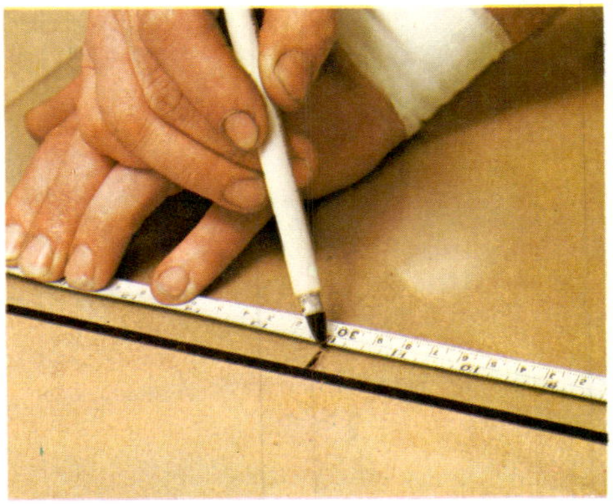
4

Mark out cutting measurements with a crayon, felt-tipped pen or chinagraph.

5

Cut glass with even-pressured strokes and always protect the wrists.

6

Use the cutter as a fulcrum or place a batten beneath the cut and press down to break.

7

Find the centre point of a circle by marking out diagonals on the area to be used.

8

Use a radius cutter with a central suction cup to mark out the circumference.

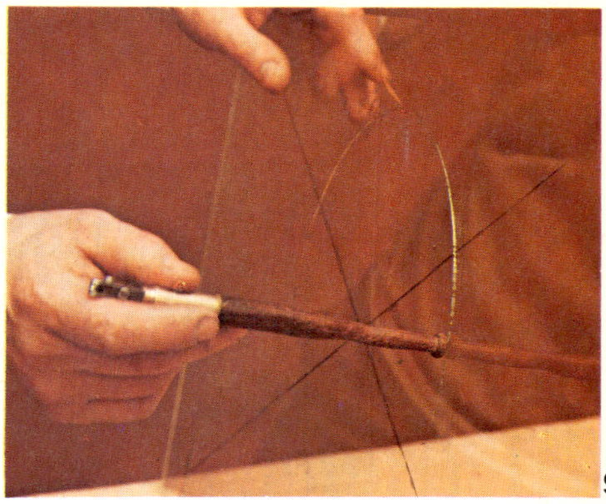

9

Tap the glass gently around the circumference to loosen and spread the cut.

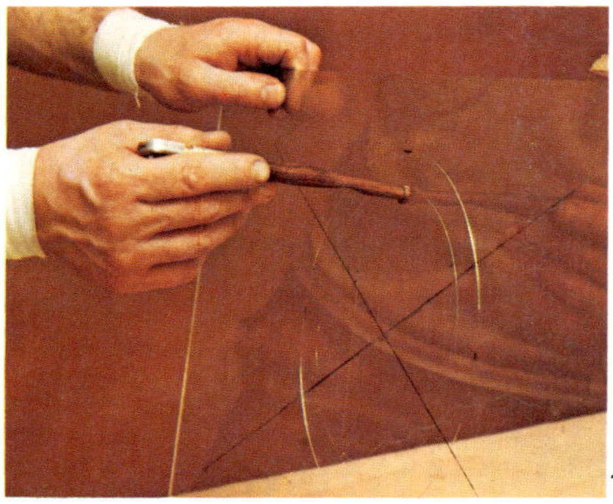

10

Cut a safety circle about ¾in. (19mm) in from the outside circle.

11

Cross hatch the centre of the innermost circle for ease in breaking out.

12

The centre can now be broken out into a receptacle by tapping with the glass cutter.

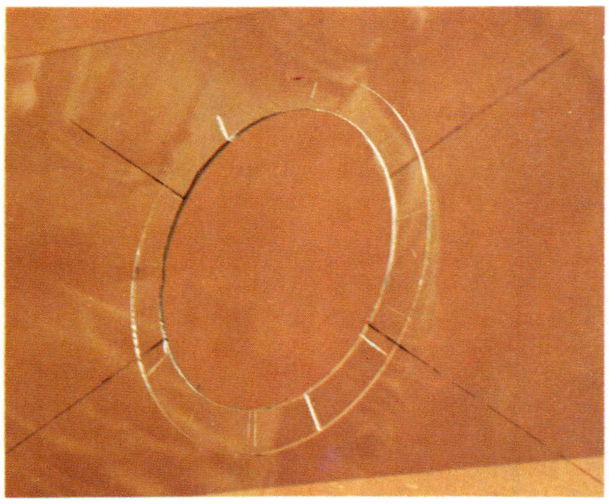

13

The inner circle has been removed and the safety ring is cross-marked for removal.

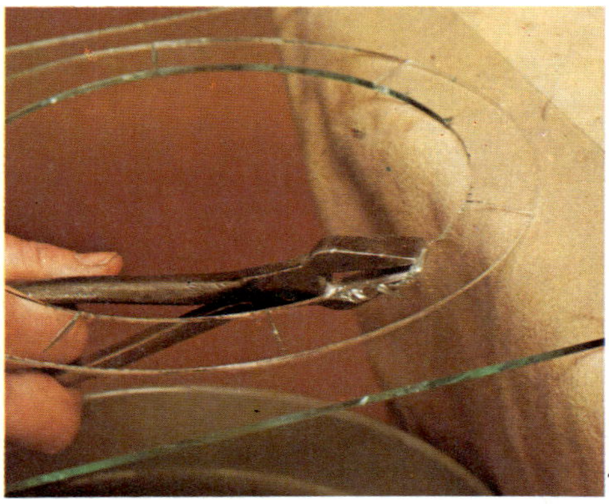

14

Remove this safety ring by gently pulling out pieces with a pair of pliers.

15

Rough edges on the finished piece can be smoothed with a wet Carborundum stone.

PICTURES COURTESY OF PILKINGTON BROTHERS LTD.

Diffuse reflection glass, commonly known as non-reflective glass, is coated on one side to reduce the amount of light reflection and thus improve viewing. It is frequently used for picture framing.

Handling glass

As long as correct methods are followed and care is exercised, glass can be handled quite safely. Experts recommend that gloves should be worn when handling glass, but they should not be too stiff, since it is necessary to be able to feel the glass to handle it with complete safety. You may also use folded newspaper or 'rubbers', such as pieces of old inner tube, as improvised 'laps' when carrying glass. Wrist bands are a further recommendation.

Always handle the glass gently but firmly; never grip it tightly. To reduce the danger of breakage, carry glass vertically and never carry it balanced on your head. When negotiating sharp corners or winding staircases, be sure to allow plenty of room for the glass trailing behind you.

When stacking glass, place it on lengths of timber which have been laid over a flat surface, and rest it at a shallow angle against a wall—25° is about right. Cushion it with newspapers or rugs and be sure that it is not kept in a place where it might be knocked over by wind.

Carrying glass sheets

One person: Small sheets may simply be tucked under the arm so that the lower edge is supported in the protected hand. Larger sheets may be carried as follows: position the centre of the bottom edge of the sheet of glass in one hand. Rest the upper portion of the sheet against the forearm and shoulder and hold the top edge of the leading sheet with the other hand.

Two persons: Large sheets can be carried as follows: The person in front should adopt the same position as descibed above for one person carrying a large sheet. The person at the rear should then support the glass with one hand cupped around the lower corner holding the bottom edge, and rest the back edge into his shoulder. The top edge is held with the other hand. The two people must be careful to walk in step at a slow pace, taking special care at changes of level and when turning corners.

While glass can be carried up ladders—supported by a second person for large sheets—this should be avoided as it can be extremely

16

To find the precise point for drilling, cross measure from the outer edges of the glass.

17

Use a spear-point bit to make a small indentation for the drill at the cross mark.

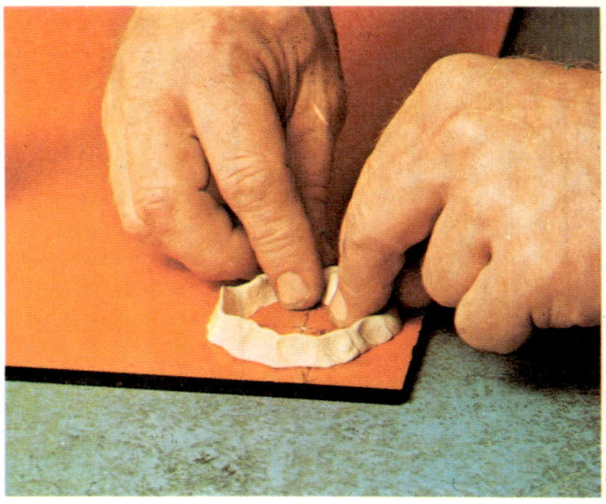
18

Build a small ring of putty around the indentation to contain the lubricant.

19

Fill the ring with lubricant and drill the hole at slow speed—here with a handbrace.

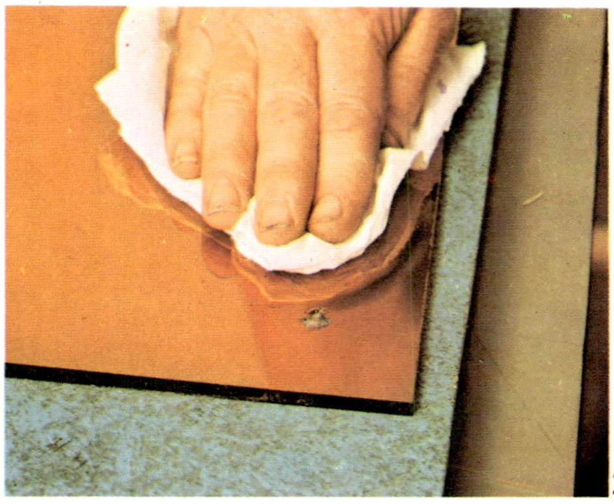
20

Just before the hole breaks through the glass, stop for a moment and clean away loose bits.

21

Finish by arrising the edges with a medium and then a fine Carborundum stone.

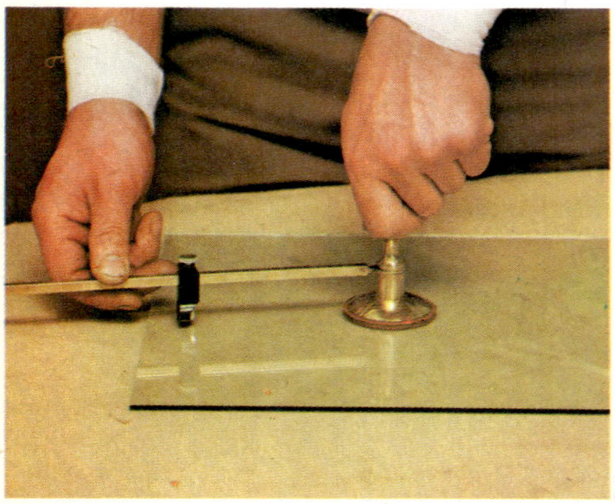
22

To cut a circle out of glass, use a radius cutter set at the appropriate distance.

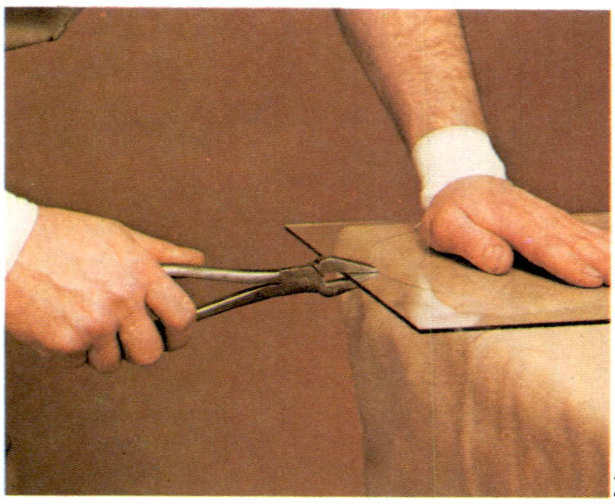
23

Spread the cut by pulling gently with a pair of pliers around the circumference.

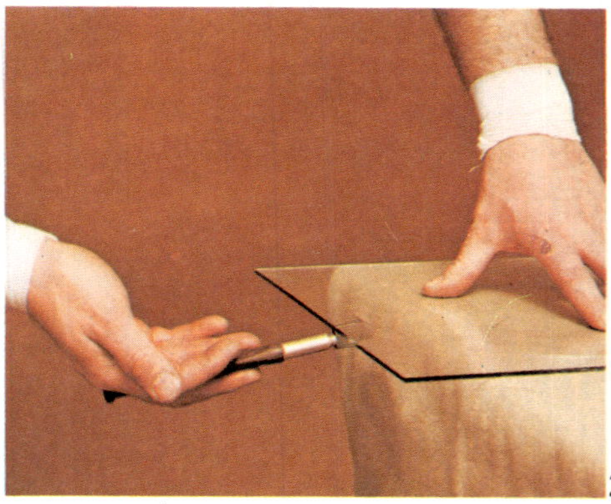
24

After using the pliers, tap along the underside of the cut with the straight cutter.

25

Work around the circle making safety cuts at each corner. Do not cross the circle.

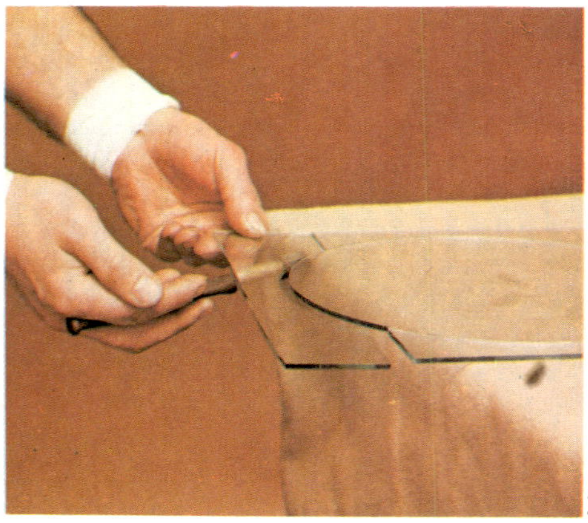
26

Tap again to loosen and spread. The piece should come cleanly away.

27

Continue working around the entire circle until you have removed all the outer pieces.

PICTURES COURTESY OF PILKINGTON BROTHERS LTD.

hazardous in the event of a strong gust of wind. If repairing a broken window pane, it is much easier and safer to reglaze it by removing the sash and doing all the reglazing at ground level. If this is not possible to do, always work on ladders with a helpmate, never alone.

Transporting glass

Glass sheets can normally be transported quite safely in a car. Small panes should be laid flat on a rear seat over a blanket. Turn the edges of the blanket just over the glass sheet, but do not cover the glass entirely—keeping it fully visible will help to prevent accidental damage. Glass carried in the boot of a car should also be wrapped in a blanket for protection. If you are carrying two or more sheets of glass, interleave them with paper or cloth.

Larger panes can be carried on a roof rack. They should be supported on a piece of blockboard at least $\frac{3}{4}$in. (19mm) thick and then covered over with a blanket. Lash the glass and the blockboard firmly to the roof rack.

Measuring and marking glass

Always double-check your measurements before cutting glass. If fitting glass for a door or window rebate, do not assume that the frames are square. Measure each side of the rebate and the top and bottom separately, noting any discrepancies between opposite sides. Deduct 2mm ($\frac{3}{32}$in.) from each of four dimensions—ie, 4mm ($\frac{5}{32}$in.) these from the total length and the total width—to find the correct cutting size. This will allow for any possible expansion of the glass in the window frame.

Preparations for cutting

Great care is needed when cutting glass. Professional glass cutters wear protective wrist bands, since a splinter of glass could cause serious injury. It is advisable, then, either to wear wristbands or to wrap your wrists with household bandages or something of the sort before cutting glass.

You will also need a large, flat surface to lay the glass on, such as a kitchen or dining table. Place the blanket on the table to protect it and the surface of the glass from being scratched.

Where odd shapes must be cut, always make a template first from a stiff card, carefully checking its accuracy. Be sure to make this $\frac{1}{8}$in. (3.2mm) smaller all round than your measurements to allow for the distance between the edge of the cutter and the cutting wheel.

A steel wheel glass cutter is adequate for nearly all types of glass cutting and is also much cheaper to buy than the traditional glass cutter's diamond. Try to choose a cutter on which the wheel is clearly visible while the tool is in use, as this will promote greater accuracy.

If you anticipate doing a large amount of cutting, you may find it useful to buy a good T-square to guide the cutter through the stroke. You will also need a ruler or steel tape for measuring. To mark the glass, a felt-tipped pen, chinagraph pencil or crayon can be used.

Straight cutting

Before doing any cutting, clean the surface of the glass by wiping it with a proprietary glass cleaner, such as Windolene, or with some methylated spirits. Mark the cutting lines on the surface and *re-check the dimensions before cutting.*

Lubricate the glass cutter beforehand by wiping it over a piece of felt which has been soaked in light machine oil. Hold the cutter so that the handle rests between the first and second fingers and the bottom of the hand remains clear of the glass. With a straight edge held $\frac{1}{8}$in. from the marking line (to allow for the thickness of the wheel), score the surface of the glass along the line with the cutter. Cut with a firm, smooth stroke, drawing your arm back while keeping the rest of your body still. Do not backtrack, as the glass may break at a point other than where you want on your cutting line.

The scoring should be completed in one operation—the object being to score the surface of the glass evenly so that the piece can be easily snapped apart. Once the score mark is made, lift the glass and tap it gently from underneath along the length of the mark. Then position the edge of a small batten, about 2in. (50mm) wide, directly under the cutting line. Place your hands on the glass surface, one on either side of the line and as close to it as possible, and press down slowly and firmly with your fingertips until you get a clean break along the cutting line.

Where you need to remove small strips or pieces of glass, score the line as before and then, using the jaws of a pair of pliers, break off the waste pieces in small bits. Do not use the notches along the outside of the cutting wheel for breaking off small segments of glass—they are intended only to be used as gauges.

Curves and angles can also be cut by scoring the glass to the shapes required from cardboard templates. Once the glass has been scored to the shape, tap it carefully from the underside and then, gripping the piece firmly on each side of the score marks, snap evenly downwards to break the glass.

Cutting circles

A special instrument called a *radius cutter* is needed to cut accurate circles in glass. This cutter is best used with the glass resting on a flat surface, although it can be used on a window which is already in position. It consists of a cutting wheel mounted on an adjustable arm which revolves on a central pivot fastened to a suction cup.

The first step in cutting out a circle is to find the centre point by marking diagonals on the piece from corner to corner. If the circle is not to be in the centre of the sheet, mark out a rectangular area where it will be and work accordingly within this area. Do this with a felt-tipped pen or crayon and then measure off the exact radius of the circle to be cut along one diagonal, beginning from the point where the diagonals cross.

Fix the suction pad of the cutter on this central point where the diagonals meet and set the arm so that the cutting edge just reaches the length of the radius. Now scribe the circle, holding the cutter firmly and applying even pressure all around. Once this is completed, move the cutting edge in about $\frac{3}{4}$in. (19mm) and scribe a second circle—this is known as a safety circle because it helps keep the edges of the glass from splintering and cutting your fingers.

Using the metal tip of an ordinary glass cutter, tap the underside of the glass upwards towards the cuts. Work slowly and carefully around both circles—the object of doing this is to open up the cuts so that the eventual removal of the waste glass is both clean and neat, without any splintered edges.

With the ordinary glass cutter, divide the safety circle into wedge sections. Cross-hatch these wedges and use the head of the cutter to tap out a small piece of glass from underneath. Carefully break out the glass in the safety circle with a pair of pliers.

Finally, tackle the outer circle. This should first be marked with a series of scores at about 1in. or 25mm intervals, but take care not to mark the glass beyond the outer circle. Break out these pieces with pliers.

Drilling glass

Normally, the only type of glass you will need to drill will be mirror glass in order to provide holes for the fixings. Special bits, made expressly for the purpose of drilling glass, should be used with a hand drill or brace. Power drills are generally not suitable (unless fitted with a speed reducer) because the speed should not exceed 350 revolutions per minute.

First, lay the glass on an absolutely flat surface. Mark the drilling position by pressing the tip of the bit on to the glass. This is to fracture the surface which will, to some extent, prevent the bit from wandering while drilling. Remember to start on the *non-reflective* side to prevent damaging the silvering. Lubricate the end of the bit with turpentine and begin drilling, keeping the drill at right angles to the surface. Do not use too much downward pressure as this will break the glass. The powdered glass thrown up by the bit will turn the turpentine white and, when this happens, more turpentine should be added. A useful tip is to make a small 'well' around the drill hole with a ring of putty to contain the lubricant.

Proceed carefully as the bit nears the other side, but do not stop turning or you will run the risk of splintering the glass around the edge and ruining the piece.

Removing rough edges

After the drilling has been completed, any rough edges can be smoothed away using two Carborundum stones—a No. 121 fine and a No. 122 medium. You will also need a natural pumice stone, some pumice powder and a wood block. Both the Carborundum and the pumice should be kept wet at all times when being used.

The first stage, called 'arrising', removes the sharp edge. Use the medium stone for this, holding it at an angle of 45°, and rubbing it downward in one direction. Once the edges have been arrised, the flat part of the edge can be ground with the face of the medium stone. Rub it up and down, keeping it in contact with the glass at all times. Follow this with the fine stone to produce a sheen. Basic polishing can now be done with the pumice stone, but to get a really fine finish, use the wood block dressed with pumice powder.

More light from interior glazing

You may want more light in your home to give it an airy, modern look, or to banish the dark corners that are found in even the best-planned houses. The best way of doing this is with interior glazing—windows set in the internal partition walls.

Windows inside the home make the best use of natural light by 'borrowing' it from outside windows and carrying it right through the whole building. There are modest and more ambitious schemes for interior glazing, but it is seldom difficult to install and, if well planned, can add new dimension to your home.

Designers and architects have tried for a long time to gain the full use of natural lighting by siting houses to face in the best direction to receive the most sun. Picture windows, fully glazed walls and roof lights have all been used in modern houses to make the living rooms bright and airy. But even where this has succeeded, the kitchen and bedrooms have often remained dingy, illuminated only by an arbitrarily placed window that lets in very little light.

The best way of overcoming this is to make the fullest possible use of natural light by installing windows in the internal walls of your home wherever they are needed. Windows in these walls transmit light but not draughts, and very little heat is lost between one room and another.

Internal windows are not a new idea. In British houses of the nineteenth century, transom windows were common. These were often in the form of a rectangular window above a doorway between one room and another. If you install transom windows, and also modernize your room doors by glazing the former solid-wood panels, you can let a lot of natural light from one closed room into another.

The hallways of many pre-1930 homes are often very dark places; in many such halls, electric light must be used even during daylight hours. A window between hall and sitting room could make all the difference here. In houses of this type, downstairs rooms are often divided into a front and a back room, which results in one room being much darker than the other.

The best way to decide how to make the most use of natural light is to sketch a plan of each floor of your home and mark on it the windows, doors and internal dividing walls, and the position of the sun at noon (due south in Britain, north in the southern hemisphere). From this, it will be obvious how natural light passes into the house and where it could be made to go to brighten a dark room.

As a result of this examination, you may wish to remove a whole wall—as many people do to make a spacious 'through-room' from the living and dining rooms. If you do not wish to undertake such an ambitious project, it is still fairly easy to install a large window between two rooms.

Planning for inside windows

A careful structural examination of your house will show which inside walls are load-bearing, that is which walls are supporting the upper floors and upstairs solid walls. Many interior walls, particularly in the upper portion of a house, are simply light, non-structural dividers built of studding and plasterboard. It is worth considering removing these altogether and replacing them with fully-glazed sliding or fixed panels. Alternatively, you can easily cut window openings in them.

Most downstairs walls in masonry houses are of brick or, more frequently, of a composition like breeze-block. If these are load-bearing, then putting a window in them will be exactly the same process as cutting and inserting a window in an outside wall. You will need a steel or concrete lintel to carry the upper masonry and floor joists for the upstairs part of the house. This is a very important structural alteration so take expert advice.

If you have central heating, interior glazing will not cause any heat loss in the house as a whole, since the temperature in one room, is more or less the same as that in the next room, and both sides of the glass are exposed to the same temperature. But the ability of glass to transmit radiant heat, called its diathermancy, will help to level out any 'hot spots' in the house.

In fact, the house may seem warmer, because daylight permeating right through the inside of a home brightens the interior and, even on a dull winter's day, this will make it seem more cheerful. Another bonus from interior glazing is that at night, one light in a room will be visible from outside all the windows on that floor which will act as a good deterrent to burglars.

Materials

For ordinary interior windows the same weight of glass as you would use for a normal outside window is adequate. But for large, floor-length panels of glass, where there is a risk of knocks from furniture and children, you should use $\frac{3}{8}$in. or 9.5mm thick float glass which can take a great deal of hard use before it chips or breaks. As a building material, glass is the cheapest of all substances used for walling, even allowing for all the other costs involved in installing it.

The glass you use in interior windows need not necessarily be ordinary clear window glass. You could use coloured, leaded glass (see the following chapter on leaded lights and coloured glass) or alternatively a decorative reeded glass. There are many different types of this translucent glass, ranging from the familiar type known as obscured glass (commonly used for bathroom and lavatory windows), to pleasingly patterned 'show' glass.

Rigid transparent plastic sheeting can be used in place of glass—or even corrugated translucent fibreglass sheeting of the type used in Britain for making patio roofs. But since some of these materials warp and discolour with age and use, make quite certain of their properties before you install them. Ask your supplier or the manufacturers for advice. These materials are cheaper and lighter than glass and, since they are more resilient, a thinner material can be used.

Interior window frames are best made from softwood, which can be either left plain or decorated with hardwood mouldings. You can, of course, use external-type window frames if you wish, but these are most usually made far too thick for interior use and the two sides of the frame (the normal inside and outside) are quite different in appearance. You are better advised to make your frames from scratch, remembering that they can be of much lighter and simpler construction than those for outside use. Your glass does not have to be put in with putty, either, since it is quite sufficient to sandwich it between two lengths of quarter round moulding. Suitable moulding shapes are available from D.I.Y. shops.

The method of putting a window into a load-bearing internal wall is exactly the same as inserting one in an external wall. Consult a D.I.Y. manual or an expert for instructions on the heavy work, and for the finer points of glazing techniques.

Getting it in proportion

Where there are stud partition walls, the best way to make good use of natural light is to remove them entirely and replace them with a large frame containing floor-to-ceiling glazing. The effect, obviously, is to give an impression of great spaciousness. You may, however, find such an open environment overpowering. You can easily lessen this effect by building trumeaux (pronounced troomo).

Trumeaux (see Fig.1) are vertical timber columns which can be used to break a large window area up into sections. The timber frame for the window and each trumeau can easily be made from stock timber sizes.

To install a window instead of a stud partition wall, or part of that wall, you should first strip off the plasterboard, remove any wiring and cut through all of the studs. The top of the window should be level with the tops of the doors in the room to help unify the appearance.

When you have cut through the studs at the top of the window, nail the top part of the frame

Right. *This opaque glass door allows a gentle natural light to filter through from the hall to the dining room. Internal glazing allows natural light to fill the whole house and brighten dark rooms.*

CONWAY PICTURE LIBRARY

ARCHITECTURAL PHOTOGRAPHIC SERVICES

Above. *Glass sliding double doors in this room add a finishing touch to modern decor. If you are glazing large areas inside the home you should use a fairly thick glass that will resist knocks.*

in place. The nails should be angled through the sides of the studs into the frame member. Then nail the bottom part of the frame to the floor.

Cut the trumeaux to length and position them between the top and bottom parts of the frame. They should be about 2ft 6in. (0.76m) apart to match the width of the doors in the room. Nail the trumeaux in place, angling the nails through the sides as before.

The next step is to nail strips of wood moulding, mitred at the corners, to the inside edges of the frame trumeaux. Nail the moulding on one side of the centre of the inside edge of the members. From the other side of the window, put the glass in place and nail strips of moulding around the frame. You do not need putty—the moulding holds the glass in position.

Left. *This attractive glass partition, designed by architect Bernard Parker of Heber-Percy Parker Perry Associates, separates the kitchen from the living areas. It prevents cooking smells from spreading to these areas and, at the same time, gives an impression of greater space within the home. It also enables you to keep an eye on the children while you are working in the kitchen.*

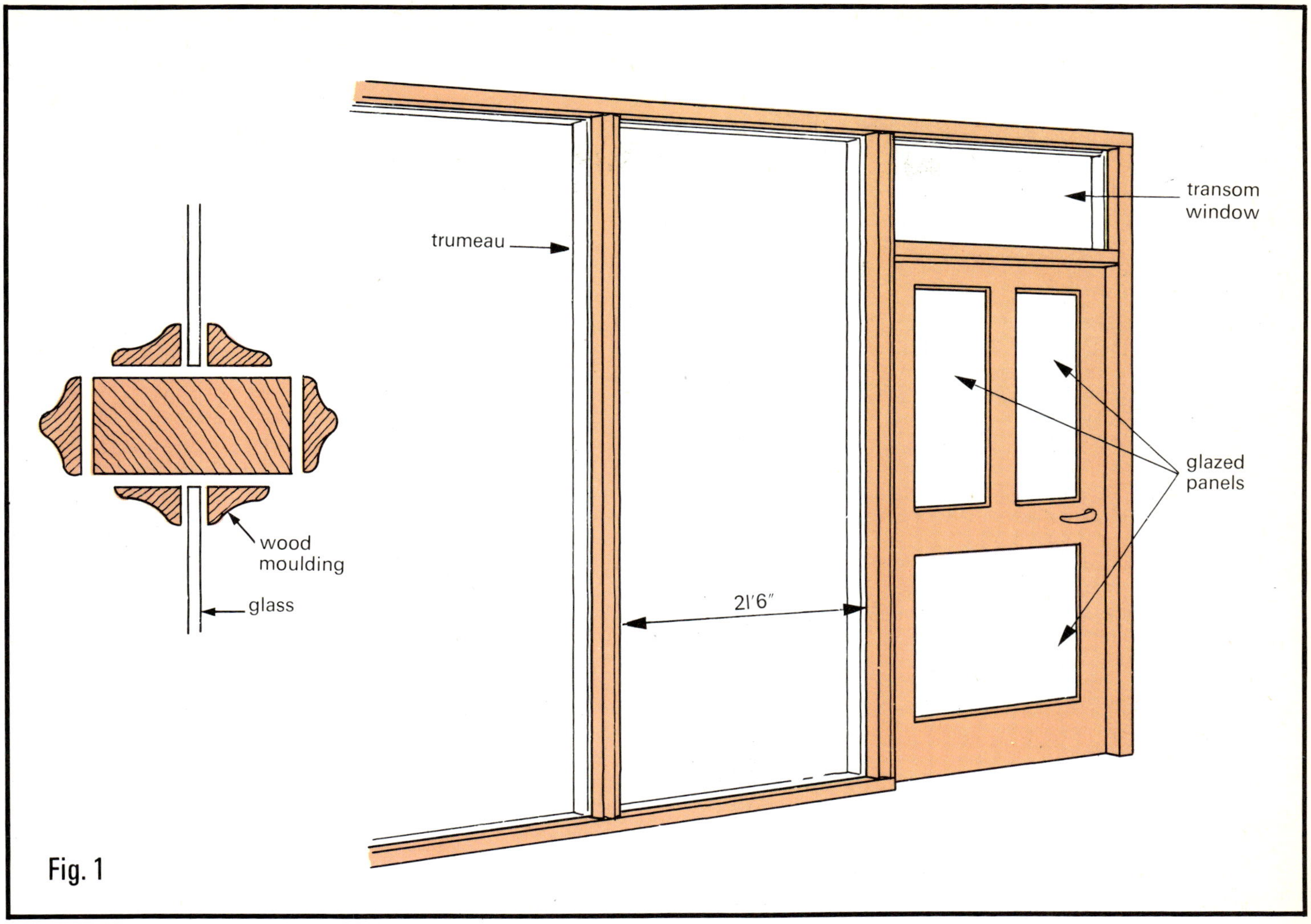

***Above.** A large area of glass can have an overpowering effect. This can be lessened by the use of trumeaux, vertical timber columns used to divide a window. These can be finished with wood moulding, as shown in the detail.*

Finish the structure by nailing decorative wood moulding, if required, to the facing surface of the frame members and trumeaux.

Wood moulding is also useful in a more modest scheme for interior glazing. If you have panelled doors in your home, you could consider removing the panels and fitting glass in their place. Use the same techniques as those for inserting window panes.

Bright ideas

Some attractive effects can be gained by painting a mural or other design on the rough side of opal glass. This is glass that has been sandblasted on one side to produce a slightly rough surface which makes the glass translucent—it lets light through but you can't see through it.

The ideal paint for this is a body colour such as gouache (a range of opaque colours ground in water and thickened with gum and honey), but you must use it fairly thick. If you dilute it with too much water the paint will not 'take' on the glass.

French enamel varnish, an alcohol-based paint, can also be used for a mural; this allows more light through than body colour. This paint is rather runny, so you should paint your design with the glass horizontal and before you put it in the frame.

If you tried to put an interior window in the party wall between your house and your neighbour's, he might take offence. But you can achieve much the same effect by making an opal-glazed frame set a short way out from the wall in a shallow box backed with matt white painted hardboard. The inside of the box can then be fitted with fluorescent tube lights; you must provide ventilation for these in the top and bottom of the box. 'Colour-matching' flourescent tubes will create much the same effect as daylight.

The job of making this fake window is very easy. First cut a piece of hardboard to the size required and paint the rough side white. Cut four side pieces of 2in. x $\frac{3}{4}$in. or 50mm x 19mm timber and drill $\frac{1}{2}$in. (13mm) diameter holes at 2in. (51mm) intervals in the pieces for the top and bottom for ventilation. Pin the pieces to the hardboard panel, nailing from the smooth, unpainted side of the hardboard. Plug and screw this to the wall, then fit the strip lights at the back, positioning their switch on one side of the frame.

From the front edge of the frame measure back the width of the glass and mark a line this distance back all around the frame. Pin thin strips of moulding along this line on the insides of the four side pieces. This is to stop the glass falling backwards when it is put in place.

The next step is to cut fascias for the edges of the frame. $1\frac{1}{2}$in. x $\frac{1}{4}$in. timber is ideal for this, though if your false window is fairly small you may wish to use narrower timber. Mitre the ends of these fascia pieces. You can also make a set of fake glazing bars to make the window look as if it had separate panes.

Get a friend to hold the glass in the box while you nail the fascia strips in place. Cut the ends of the 'glazing bars' to fit against the strips and fasten them in place with glue and a nail through each end.

To complete the illusion, and also to hide the edges of the box, treat this false window like an ordinary window, putting up curtains and a pelmet. With its own interior lighting, it then becomes an attractive and eye-catching light source.

Glass bricks were once used to aid natural lighting inside the home, particularly in the bathroom, where they were used to make the walls of shower cubicles. They are, however, very expensive, very heavy and take up a lot of room. Where you do require privacy, in the bathroom or elsewhere, two thicknesses of reeded glass, preferably with the patterns running at right angles to each other, or a sheet of opal glass and a sheet of reeded or blistered glass provide a much better alternative.

These materials can also be used to create an attractive splashback behind the bath in an untiled bathroom. They should be fastened to the wall with plugs and mirror screws.

Leaded lights and coloured glass

There are many hundred of thousands of houses, particularly British ones built before 1940, which have leaded windows. In really old houses, these may be formed of small squares of glass, but diamond-shaped pieces are more usual in newer buildings. Damage to a leaded window often causes the owner to throw it out and replace it with a sheet of uninteresting clear glass. The same applies to the coloured glass windows often found in Victorian and Edwardian houses. But here, instructions are given for the repair of leaded glass windows, and for making your own coloured-glass window panels. A coloured-glass window could add interest to any home—and the technique can be adapted to making fire screens and other items.

A leaded window depends for its strength on special channel-sectioned strips of lead into which the pieces of glass are slotted. These are called *calms* (pronounced either *cams* or *cames*) and are produced in a variety of different widths and channel thicknesses. The calms are cut into proper lengths and soldered together at each intersection so that the separate panels of glass are each fully surrounded by lead strip on all sides. The joins between glass and lead are then made waterproof with a special cement. The complete leaded window is remarkably rigid and quite weatherproof.

Materials

Lead calms can be bought through most glaziers; if they do not stock them, at least they should be able to tell you where you can get them. Two types are available. One has a flat surface, usually with a slightly beaded edge, and the other has a slightly convex surface. Both types are symmetrical in cross-section. The second type is stronger, and probably better suited to normal leaded window work, but it is less flexible than the flat type. For repair work, your choice will be governed by the type of calms originally used on the window.

The special cement used for sealing the glass to the lead can easily be made at home. Begin by mixing equal parts of plaster of paris and powdered whitening, and add to this one part of lamp black and about half a part of white lead. Now dissolve the mixture in boiled linseed oil until it takes on the consistency of very thick treacle, then stir in half a part of red lead. Finally, add turpentine or turps substitute slowly, stirring the mixture until it attains a thin, creamy consistency. This cement is inflammable; it is also poisonous because of the red and white lead in it. It is best when handling it to wear gloves—and you must make sure you wash off any that gets on your skin and make certain that none contaminates food.

A polishing compound is used on the lead of the completed window to make it a decorative shiny black. This used to be done with black stove polish, and if you can still buy this in your area, use it. Otherwise, you can make up a suitable mixture by mixing vegetable black and turps substitute into a thick paste rather like shoe polish.

Tools

Making and repairing leaded windows is an old craft, and the tools required are simple. First, there is the glass cutter. If you are only working with glass having straight edges, a diamond cutter will suffice (this has a very small, cheap, industrial diamond and is not at all expensive). But if you are cutting curves, then a wheel cutter is the only type to use. For cutting the calms, you will need to buy a stripping knife. Cut the blade down so that it is about 2in. (50mm) long and quite stiff—the easiest way is to snap the blade approximately and then grind it to shape—and then re-sharpen the cut end. This rigid knife is an ideal tool for shearing through lead. A $\frac{1}{2}$lb or 200g hammer is used for tacking against the glass on assembly (described below); its handle doubles as a tapper to force the glass into the channel of the calm. An oyster knife is used for prizing open and straightening the flanges of the calms. Use this exact type of knife if possible; otherwise use a knife with a large handle, very thick and rigid short blade, and a rather blunt edge on both sides. A flat piece of boxwood, sometimes called a *lathekin,* is needed for smoothing down the flanges. A parallel-sided palette knife or thin boxwood paper knife should be used under the flanges when they are pressed down so as to leave a small gap for the cement to be run in.

For minor repair work or making one or two small panels for windows, an ordinary soldering iron can be used. This may be either of the solid copper type requiring a flame for heating, or a largish electric one. But if you want to take up this work as a serious hobby, you should buy a gas soldering iron, which runs off the domestic supply or bottled propane gas. This type of iron is made with a circular-section copper bit set at right-angles to the handle, and is the easiest type of all to work with. The solder used in leadwork of this kind is the type known as tinman's solder. This should be bought, if possible, $\frac{1}{16}$in. or 1mm thick. Glaziers' suppliers sell these narrow strips. Use a tallow candle as flux.

If you are attempting stained-glass work, or work involving the cutting of intricate shapes of glass, then you will need several pairs of special pliers called *grozers*. Two main types of grozing pliers are useful. The first has smooth, wide jaws and is used to snap off long, narrow strips of glass after they have been scored with a glass cutter. The second pair closes to a V-shape; it is used to chew the edge off a piece of glass where the amount to be removed is too small for the cutter to be effective. You can also use grozers to chew tight inside curves, bit by bit. Both types of grozers can be made from ordinary pliers by softening them in a fire and then grinding or filing their jaws. Alternatively, glaziers' suppliers may stock them, or be able to tell you where to find them.

Repairing a broken lead light

It is possible to make simple repairs to a leaded window when it is in place. You will, however, find it easier as well as safer to remove the *light* (this is the complete lead-framed assembly of glass) from its frame and work on a flat, clean workbench. If this cannot be done, and if it is also impossible to unscrew the window hinges, then consult Figs.12-16 to show you how to work on a window in place.

To remove the old light, proceed in exactly the same way as for the removal of a normal single pane of glass. The light will probably be secured with tacks and putty into a rebate in the window frame. Rake out the old putty, draw out the tacks with an ordinary pair of pliers, and then carefully ease out the light. You may find it necessary to hold a strip of wood along the inside edge of the lead and tap this lightly with a hammer in order to free the lead from the window frame.

The first operation in replacing a damaged pane is to turn back the flange of the lead calm on one side. Which side you choose depends on which is the most convenient side for you to work. Using either the oyster knife or the palette knife, gently insert the end of the blade between lead and glass at a point approximately in the centre of one side of the pane, and slide the knife back and forth with the handle as close to the glass as possible. This will slightly stretch the lead, and at the same time rake out the cement. Any attempt to open out the lead quickly, using the knife in the same way as you would use a screwdriver to open a can, will damage the lead. You must work gradually, progressively raising the handle of the knife on each pass so that the lead opens up smoothly.

You will find it impossible to open up the corners properly because of the soldered joints at the intersections of the calms. Turn the stripping knife so that the edge of the blade lies against the glass and the sharpened end points into the corner, then push the sharp end into the lead joint, cutting it more or less down the centre of the angle formed by the joined calms. Do this at all four corners of the piece, then continue opening up the leads until they are bent out at right angles.

If the damaged pane of glass is just cracked, you may have difficulty in removing the pieces, particularly if it is still more or less intact. Here,

Using the basic techniques described here you can make an attractive coloured glass fire screen like this. The design is on page 70.

you may find it easier to score the broken pieces with the glass cutter, and then break out the glass in small bits. Hold the glass with pliers and twist them to break it. Guard against splinters breaking off from the edges as you twist, as these can fly into your face. Wearing a pair of lightweight goggles—as you should when grinding metal—is the best precaution.

Once all the old glass is out, make sure that the lead flanges of the calms are neatly bent straight out, then make a rubbing of the opening, using a piece of paper and the side of a pencil as shown in Fig.13. Cut a fresh piece of glass, using the rubbing as a pattern. You will find that even though you have cut into the calms at the corners, you will still have difficulty in opening the flanges right back at this point. To get the glass to fit, therefore, you must gently chip off the pointed corners; this is done with the grozers as shown in Fig.14.

Hold the glass in one hand, take a grip of the pointed corner with the grozers, and squeeze them gently until a small piece breaks off the corner. Nibble gently at the glass bit by bit until you have removed enough. Once the corners have been rounded off in this manner, the new piece of glass should fit neatly into the leads. Note that this process of removing the corners of the glass is only necessary when you are replacing glass in an existing leaded panel and is not needed if you are assembling a panel from scratch as described below.

Hold the palette knife against the glass and gently rub the lead down on to it so that there is an even, parallel gap between lead and glass equal to the thickness of the knife. This should be done with the boxwood piece, or lathekin, which should be rubbed along the lead to press it down to the blade of the knife. Again, work gently and progressively up and down the lead, moving the knife and the lathekin together. The joins in the calms, where you have cut into the thicker, soldered lead, may be hard to press down in this way. If you have any difficulty, gently tap the lead with the flat side of a hammer, supporting the other side of the glass with your hand.

Make sure that the gap between glass and lead flange is about $\frac{1}{16}$in. (1.5mm) and that it is quite even all the way round. For this type of repair, it is not necessary to re-solder the cut joins in the calms, particularly if you have made them on the inside of the window. Nor is it necessary to use special cement on the glass; an adequate repair can be completed simply by puttying the new piece. Mix up a small amount of linseed oil-based putty with lampblack to produce an evenly black compound, and press this into the crack between glass and lead all the way round on one side. Now do the same on the other side. The action of forcing the putty between glass and lead on the second side will push the glass to a more or less central position in the calm and give an equal thickness of putty on each side. Clean off excess putty with the palette knife.

NIGEL MESSETT

Stained-glass work

The term 'stained-glass' is generally applied to any kind of coloured glass in leaded windows. It is, however, more properly clear or coloured glass into which coloured metallic oxides have been fired in a kiln. In this way, a piece of glass which is basically one colour can take on a number of different hues, and so form a portion of a transparent painting of the type found in church windows. Proper stained glass work of this kind is quite within the capabilities of a dedicated amateur who can afford the necessary equipment, and is described in specialist books on the subject.

A far easier technique for the average amateur is making a leaded window with ready-coloured glass. A panel of leaded coloured glass can be used in a hallway or bathroom window (where the daylight coming through will produce exciting effects) or can make an attractive and unusual firescreen as described here. The actual method of making up the leaded glass remains the same whatever its final purpose.

If you are going to make your own design, there are several points to watch. There are seven usual thicknesses of lead calm ranging from $\frac{5}{8}$in. (15mm) down to $\frac{1}{8}$in. (3mm), but the most easily obtainable sizes are probably $\frac{1}{2}$in. (13mm), $\frac{5}{16}$in. (7.5mm) and $\frac{1}{4}$in. (6mm). In drawing up your design, you must allow for the overall width of the lead strip, which has an important effect on the appearance of the panel, and also for the thickness of the calm core, normally about $\frac{1}{16}$in. (1.5mm), which controls the size of each piece of glass. For example, it is pointless to use a $\frac{1}{2}$in. (13mm) wide strip of glass between two $\frac{1}{2}$in. (13mm) calms; the glass will not be visible.

In any case, very small pieces of glass, or pieces with intricate shapes or sharp inside curves should be avoided. You *can* groze out very small pieces with sharp inside curves that are too small to cut with the glass cutter, but there is a practical limit of about $\frac{1}{2}$in. (13mm) internal radius. Internal angles (rather than curves) cannot be cut at all because the glass would almost certainly break. Where an internal angle is required in a glass piece of one colour, the shape must be made out of two pieces leaded together. It is also inadvisable to try to shape very narrow curved pieces of glass, since this requires great skill to avoid breakage.

Fig.1 shows a design for a coloured window or screen, which you can follow by scaling up the dimensions on to a large sheet of white paper. Alternatively, of course, you can design your own. First, find out what colours of glass are available in your area. This should give you some idea to start you off, and you can then make rough sketches until you arrive at a shape and combination of colours which pleases you. Scale this up to full size, following the instructions given below.

The cartoon

From your rough sketch, the next stage is to produce a full-sized working drawing sufficiently detailed to enable you to cut your glass to the exact size and shape you need. This drawing is called a *cartoon;* you can draw it on tracing paper or any white paper. Take particular care in enlarging the sketch to the cartoon stage, and mark all the lines with a thick pencil or felt-tipped pen that draws a line about $\frac{1}{16}$in. (1.5mm) wide. The thickness of this line represents the thickness of the central core of the calms. The actual cutting line of each piece of glass will then be on the near edge of the line next to it. Failure to allow for this, or any inaccuracy in cutting the glass to fit *inside* the lines, can produce an alarming change in the final size of the panel.

Cutting glass

Use a cutter with a new wheel and practice on odd scraps of glass to learn the right way to hold and use the tool. Exert enough pressure to score the surface of the glass without the wheel skidding across the surface. Don't press so hard that the wheel bites in and chips the surface of the glass. Lay the glass you are cutting from over the cartoon, and hold the cutter in such a way that you can make a clear continuous stroke without faltering. Dipping the cutting wheel into paraffin before each cut lubricates the wheel and helps avoid chipping.

In making straight cuts on glass lying on the bench, always cut towards you, but stand slightly to one side so that your elbow will not hit your body as you move your arm. To cut a circle or a curved line, it is best to hold the cutter firmly and push it away from you, following the line on the cartoon. This is why a diamond is no use for cutting curves: the wheel is much more positive in its movement.

Cut out all the glass for your leaded panel at one go, ticking off each piece on the cartoon so as not to get confused over which pieces are cut and which are not. Lay the cut glass out on a piece of clear glass or hardboard as you cut it out until you have built up the whole picture. This way you can see what you are doing and correct any mistakes as they occur.

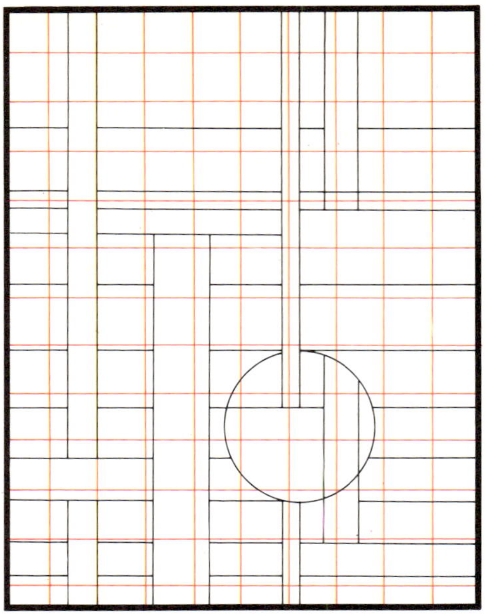

1

Fig.1. The design for the firescreen panel shown on the preceding page. The squares are 2in. (50mm) and can be scaled up from this.
Figs.2, 3 & 4. Alternative suggestions for making attractive coloured glass panels.
Fig.5. The proper way to hold the wheel cutter for cutting out the pieces of glass.
Fig.6. How to cut the calm using the sharp edge of the knife. Press straight down through the lead and the flanges will not close up.
Fig.7. Assemble the pieces systematically on the cartoon and mark each lead for cutting. See how the first corner of the design is squared up against two wood battens at right angles. The first pieces of coloured glass are being assembled on top of the cartoon.
Fig.8. The circular detail is contained in one strip of lead bent round the glass. The pieces already fitted are held in place with short lengths of calm nailed to the bench top.
Fig.9. Opening out the edge calm flanges.
Fig.10. The complete design trued up square and ready for soldering the joints on one side.
Fig.11. The gas soldering iron being used to solder up. A tallow candle is rubbed on each calm joint as flux. Complete one side at a time.

2

3

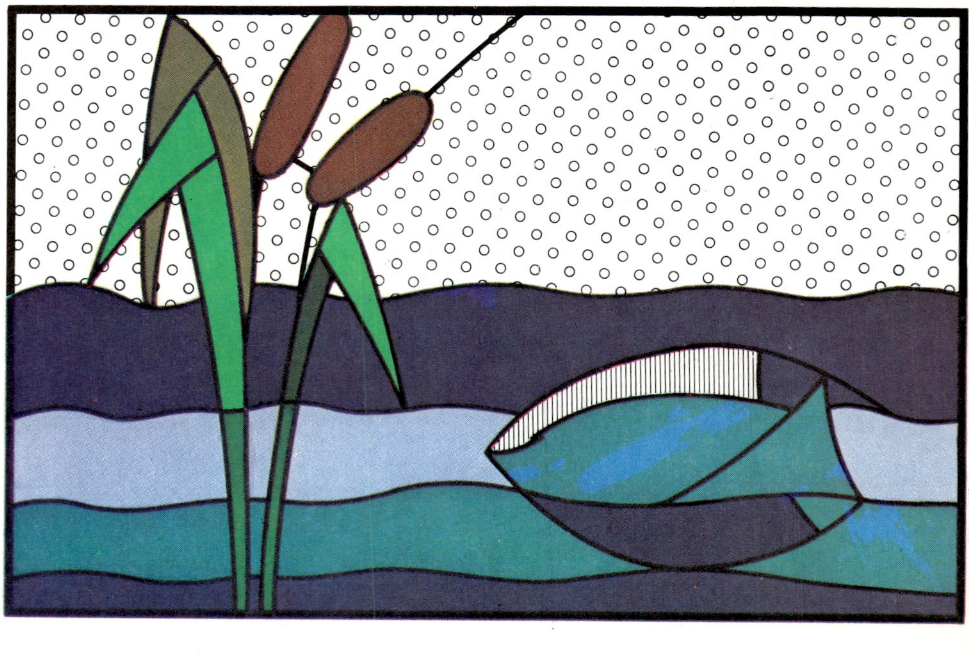

4

5

6

7

8

9

10

11

Preparations for work

Your work area should be clear and absolutely smooth—a sheet of plywood or hardboard laid on a work-bench will make an ideal surface. Pin the cartoon to this and tack two strips of wood about $1\frac{1}{2}$in. x $\frac{1}{2}$in. (40mm x 13mm) in cross-section, and slightly longer than the sides of the panel you are making, to two adjacent sides of the picture at right angles to each other. These strips act as stops to hold the panel steady as you build it up.

Measure up the total length of each size of calm you will require and add about 10 per cent to allow for inaccuracy and wastage. Stretch each length of calm slightly by holding one end in the vice and pulling gently on the other with a pair of pliers. This will straighten out any kinks and twists.

Cut off a length of your thickest calm to make one side of the panel; take this measurement off the cartoon and add about $\frac{1}{2}$in. (13mm) for safety. Place this piece of calm against one of the two laths you have tacked to the bench and cut another length of calm for the second side. There is no need to mitre the corners: they can just be butt jointed. The calm should be cut by pushing down on its face with the sharpened stripping knife. If the knife is sharp enough and you exert enough force, it will cut through the lead cleanly without closing up the channels. Support the lead on a wood block for cutting.

You are now ready to start assembly. Place the first piece of glass—the one which fits into the corner created by the two lengths of calm—firmly into place and make sure that it has slotted into the channels properly. Now take a length of calm and cut it to fit along the exposed side of this piece, butting against the first piece of calm at the end. Note that you must cut this length of calm shorter than the true length of the side of the piece of glass, so as to allow for the width of the flange of the lengths of calm it lies against. This stage is shown in Fig.7.

Keep adding pieces of glass and lengths of calm until they stretch all the way along one side. Make quite sure that the pieces of glass are properly inserted into their slots; you will have warning of any problems here by comparing the glass with the cartoon upon which it is superimposed. Tap the exposed edges of the glass pieces all along their length with the wooden handle of the hammer to ensure that they are all a snug fit. To prevent the pieces from moving, cut a very short piece of calm about $\frac{1}{2}$in. (13mm) long and slot it over the end of the last piece of glass. Then drive a short nail into the bench hard against the scrap piece of calm to hold it in place. These can later be removed as work proceeds.

You can now continue with the next row of glass pieces in the same manner. Lines which are supposed to be straight (as shown on the cartoon) should be represented by long, straight pieces of calm, and you must see that these lengths are in fact straight, and that any parallel lines are really parallel. Check this as you go, but there will also be a final check before soldering up.

Any curved pieces of glass, such as the circle shown in the design in the illustrations, must be framed in curved calm. This is easily formed by drawing the lead through the hands and gently pressing it in to shape. It bends very easily.

Once all the glass pieces are in place with their separating calms, you can cut and fit the final straight lengths of thick calm to the remaining two sides. Check the panel for squareness by measuring the sides and diagonals. An ideal way of ensuring that the edge calms are kept straight is to cut a long strip of plain glass, slot it into the open outer side of the calm and hold it down on the bench with tacks along the opposite edge. Make a final check to see that the pieces of calm in the design are laid straight and in their proper position. The thickness of the glass is slightly less than the size of the gap between the calm flanges, so some movement can easily take place; it can also just as easily be corrected. The panel is now ready for soldering up. This must be done before removing it from the bench.

Soldering

Clean the end of the copper bit of your gas soldering iron. This must be absolutely bright and untarnished, or you will not be able to tin it properly. Light the gas and regulate the heat until the iron is at the correct temperature. Achieving and maintaining this exact temperature take a fair amount of practice. If the iron gets too hot, you will not be able to tin it, and furthermore, if it overheats in use you will burn away the lead calms instead of joining them.

Test for the proper heat during the tinning process by melting a little solder in a tin lid, letting it cool and rubbing the end of the iron on this as the bit heats up. When the solder starts to flow, you have reached the right temperature. A little powdered resin (rosin to some people) will help the solder to flow if scattered lightly over the solder in the tin lid before you start tinning the iron.

Once the copper bit is properly tinned—a smooth 'puddle' of tin should adhere evenly to the end—you are ready to solder. Rub the tallow candle over each calm joint. Now hold the thin stick of solder in one hand and the iron in the other. Melt off about $\frac{1}{8}$in. (1.5mm) of solder and quickly touch this on the calm joint. With a little practice, you will find that each joint can be soldered in this way in a matter of seconds. Do not press the iron on to the calm—the lead will instantly dissolve. The technique is to touch the joint with the puddle of molten solder on the end of the iron—and it really should be just a touch—the heat of the iron passing into the lead calm through the molten solder.

When you have completed one side, the panel will be rigid enough to be removed from the bench and turned over. Solder the other side, then give both sides a good scrub with a dry stiff-bristled brush and you are now ready for the next stage—cementing.

Cementing

Cementing is rather a messy job, so spread old newspapers over your bench and wear an old apron to protect your clothes. It is advisable to wear thin plastic or rubber gloves, since the red and white lead in the cement is poisonous and could cause serious trouble if you have any open cuts on your hands. And after all that glass cutting, you will probably have one or two cuts.

Spread the cement all over the panel with a scrubbing brush, rubbing it in everywhere until every crevice is filled. Now spread whitening or plaster of paris all over the panel. This will speed up the drying of the cement. Turn the panel over and repeat the procedure for the other side. Leave the panel with both sides whitened for an hour or so, then brush off all the whitening with another scrubbing brush. Any scum left on the glass can be wiped off with a cloth soaked in turpentine or turps substitute. Do this before the cement has time to set really hard, or it will be difficult to remove. Now leave the panel for a day or so for the cement to harden thoroughly.

To prevent your cement brush from setting solid, clean it immediately after use with turps substitute and then rinse it in hot water and detergent.

Complete the job by giving the panel another stiff brushing, this time with stove polish or the substitute described under the heading 'Materials'. This will shine up the lead calms, and also cover the cement joints with a protective coating.

Installing the panel

If your leaded glass panel is to go into a window frame, it can be fitted in precisely the same way as you would fit a normal one-piece clear glass window, this is a relatively easy task. First of all, though, the open edges of the calms at the sides of the panel must be closed up by folding one edge over the other.

You may wish to use your panel as a decorative firescreen like the one on page 69. To do this, make a light framework of brass or aluminium into which the panel will fit. Alternatively, you could mount the panel in a plain wooden frame, or even in a decorative one made from picture-frame moulding. The panel needs to be lit from behind to look effective and so should be hung in front of a window or a light and not just placed on a wall like a picture.

12

13

14

15

16

Fig.12. *Opening out the lead flanges around the broken piece of glass. You may have to cut the cracked section into smaller bits in order to remove them from the window using pliers.*
Fig.13. *With the lead flanges standing out at right-angles and the old glass removed, take a rubbing on a piece of paper and cut the new sheet of glass to suit this pattern.*
Fig.14. *Remove the points of the corners by grozing the glass with special grozing pliers.*
Fig.15. *Place the new piece of glass in the prepared opening and carefully rub down the flanges of the calms with a boxwood lathekin.*
Fig.16. *Seal both sides using linseed putty mixed with lamp-black. Clean off excess putty with a sharp knife and after it has hardened for a few days, polish it with black stove polish and shine the lead with a soft cloth.*

Leading lights

NIGEL MESSETT

Modern rooms need modern lighting to make them look their best. Too often, a room with stylish contemporary furniture is marred by dull, old-fashioned lamps. But these acrylic lights, with their brash and uninhibited use of shape and colour, will add interest to any room.

Two of the lights are designed as table lamps, and the third for fixing to the ceiling. The table lamps are set on simple wooden bases that require little knowledge of carpentry to make. Otherwise, all you need to make the lamps is a small amount of sheet acrylic, a few proprietary light fittings and (for the ceiling light only) 150 yards or metres of clear nylon fishing line or thick thread.

You should read the detailed instructions on cutting and glueing acrylic sheet before beginning any of the work on these lights. In British countries this sheeting is better known by the brand name Perspex.

The box table lamp

This is the simplest of the three lights to build, since all its acrylic parts are plain squares or rectangles. The white sphere in the middle is an ordinary white glass globe of the type used for kitchen and bathroom ceiling lights.

To build this light you will need:
a piece of 'bronze' acrylic (which is actually smoke-grey) measuring 28in. x 19in. x $\frac{1}{8}$in. (712 x 432 x 3mm),
an 8in. (203mm) diameter white light globe,
a 'batten' light fitting (for fixing a bulb directly to a wall or ceiling) incorporating a globe support,
as much two-core electric cable as you need for the flex,
a 40 or 60-watt bulb (no larger),
and a 9in. x 9in. x 1in. (229 x 229 x 25mm) wood base (see below).

Construction is very simple. Cut the acrylic (without removing the protective paper) into two pieces 9in. x 9in. (229 x 229mm), two 9in. x 8$\frac{3}{4}$in. (229 x 223mm), and one 8$\frac{3}{4}$in. x 8$\frac{3}{4}$in. (223 x 223). Clean and polish all the edges as accurately as possible.

The smallest of these pieces (8$\frac{3}{4}$in. x 8$\frac{3}{4}$in.; 223 x 223mm) is the base, and will need some holes drilled in it (see the exploded view, Fig.1). Examine your light fitting and measure the diameter of the external threaded part of the bulbholder, which will pass through the base. This measurement will normally be 1$\frac{1}{8}$in. (28mm), but some light fittings are larger or smaller.

Examine also the part of the fitting that supports the glass globe. It will probably have a ring of holes in it for ventilation. Measure the diameter of the ring.

Now take the acrylic base piece (still without removing the paper) and find the exact centre of the square. Set a pair of compasses to the *radius* of the ring of ventilation holes in the globe support (i.e. half the diameter). Stick the spike of the compasses in the centre of the acrylic base piece and draw a circle.

Drill a small pilot hole (say $\frac{1}{8}$ in. or 3mm through the centre of the base, try to make the measurement accurate). Then enlarge the hole to 1$\frac{1}{8}$ in. or the diameter of the shaft of your light fitting. The best types of drill to use for this are *either* a hole saw of the correct size *or* any type of large wood bit with a central spike and smaller spikes on the outer edges of the cutters. Whichever type you use, drill slowly and carefully halfway through, turn the acrylic over and drill the rest of the way from the other side. You could also cut the large hole with a fretsaw.

Now drill a ring of ventilation holes around the circle you have drawn. Six $\frac{1}{4}$in. (6mm) holes is enough. They will be visible from certain angles, so drill them neatly, at equal spacing, and preferably lining up with the existing ventilation holes in the globe support.

Remove the protective paper from both sides of all the pieces of acrylic. Then take the base piece and stick cellulose tape (e.g., 'Sellotape') to one side of it all round the edge and hanging over it, so that the piece is edged by a rim of tape. Stick the tape to the flat surface, not to the edge of the sheet.

Lay the base down with the sticky side of the tape uppermost. Take one of the two 9in. x 8$\frac{3}{4}$in. (229 x 223mm) pieces of acrylic and hold it so that the long sides are vertical. Then lower it on to the tape along one side of the base, press it down to hold it firmly and fold the remaining width of that strip of tape upwards, so that the acrylic side piece is firmly taped to the base. Repeat on the opposite side of the base with the other piece. Then take the two 9in. x 9in. (229 x 229) pieces and tape them to the base in the same way. You will now have an open-topped box. Tape the four sides together firmly along their outside edges.

Stick the box together permanently by running chloroform down the insides of the joints. (Be sure you work in a well-ventilated room, with no naked lights.) This is done by tilting the box so that the joint being glued forms a gently-sloping trough, and dribbling a very little chloroform in from the top. Take care not to let the chloroform run anywhere it is not wanted; it attacks the surface of acrylic. Allow each joint to dry (this only takes about five minutes) before going on to the next.

When all the glued joints are dry, remove the tape and clean off the sticky traces it leaves with methylated spirit. Then assemble the lamp as shown in the exploded view. Instructions for making the wooden base are given below.

The slotted table lamp

This is also made of $\frac{1}{8}$in. (3mm) acrylic sheet. A dark colour, such as the blue shown in the photograph, gives the best results. You will need a piece of acrylic 3ft x 2ft (915 x 610mm), a 'batten' light fitting with bulb, wiring and a wooden base as described below.

The shapes to cut out are shown in *red* on the cutting diagram (Fig.2). This diagram is exactly half size, so draw full-size versions yourself, preferably on graph paper for maximum accuracy, and use them as templates.

Cut thirty of piece A, two of piece B, two of piece C and one of piece D. Just cut out the basic rectangles; do not cut slots or drill holes yet.

Stick the thirty pieces A together face-to-face with double-sided tape so that all the sides and ends are exactly level. You will have to remove

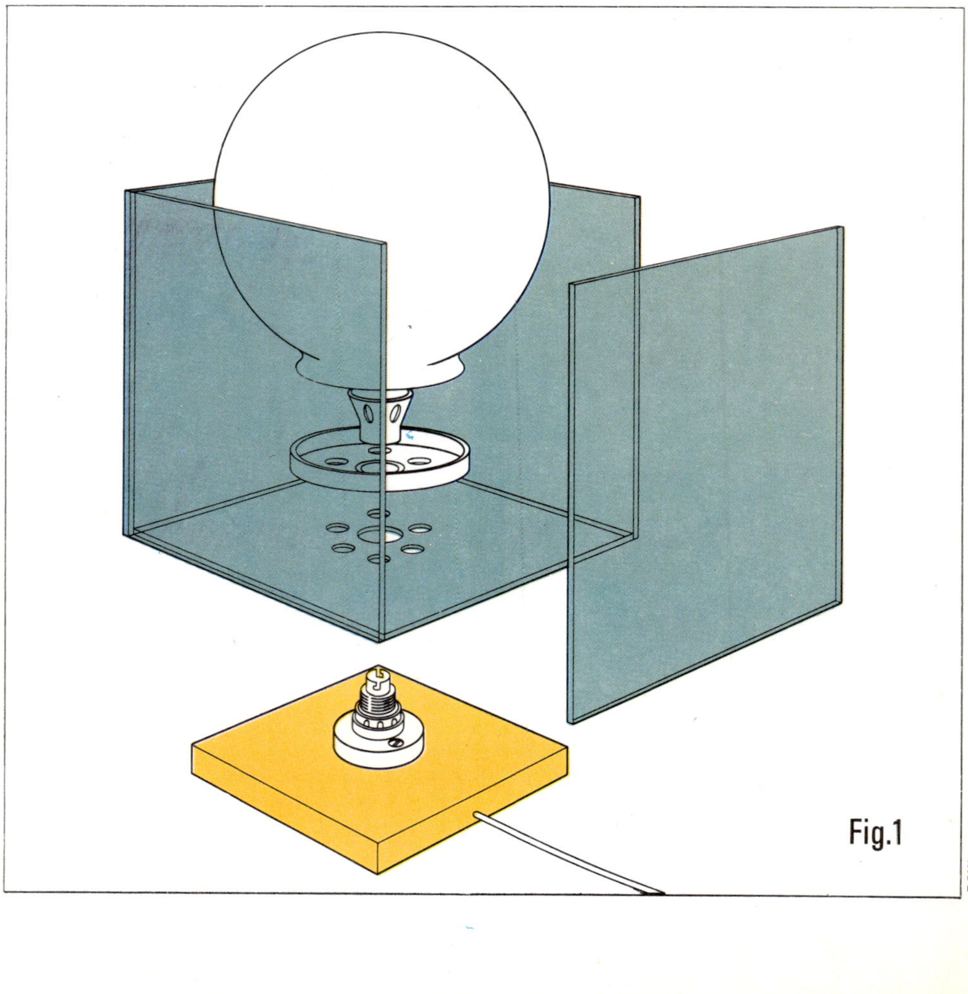

***Fig.1.** (right). An exploded view of the box table lamp shown on the opposite page. Note the ingenious use of a standard ceiling light globe, which is installed upside down.*

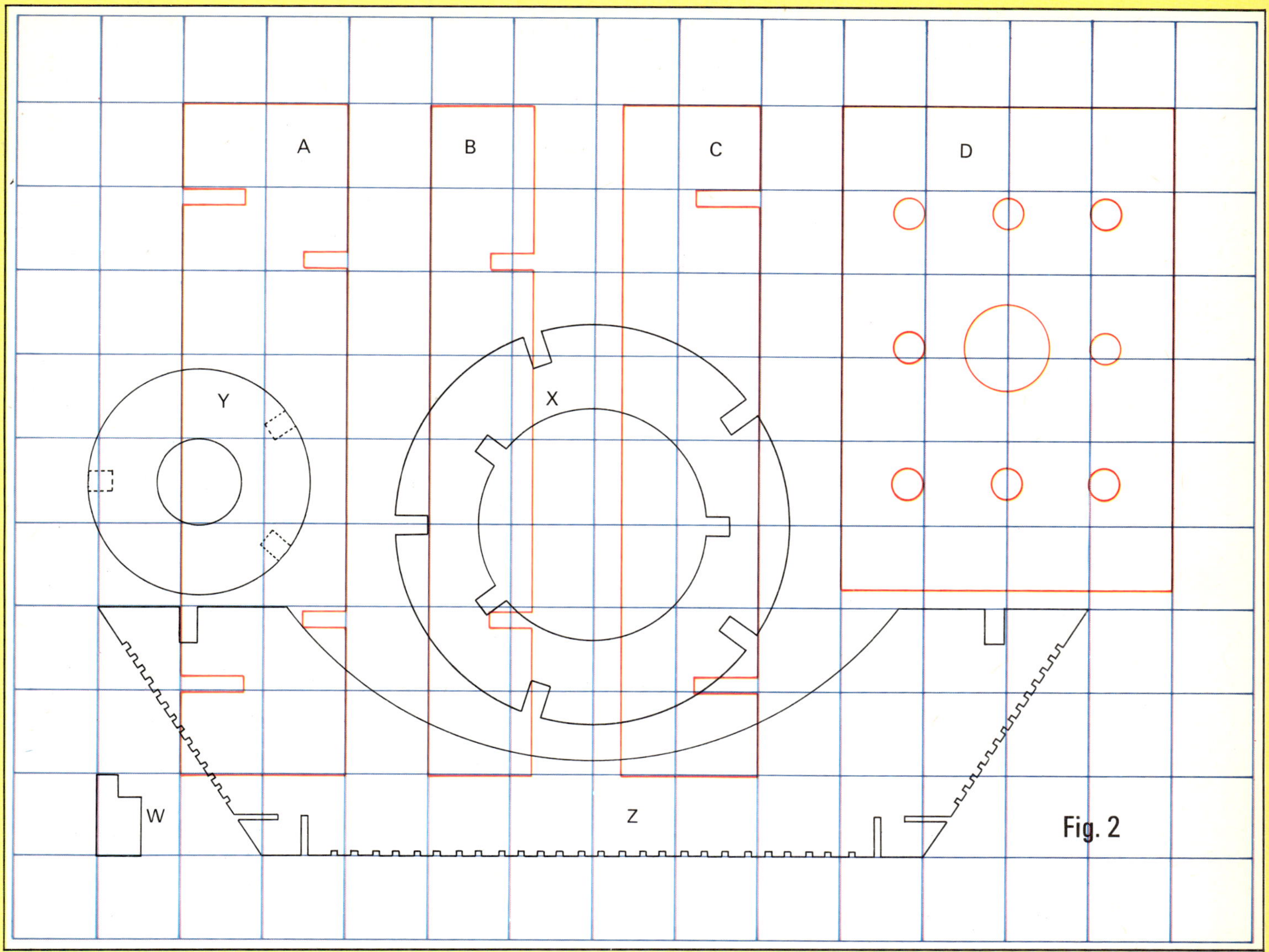

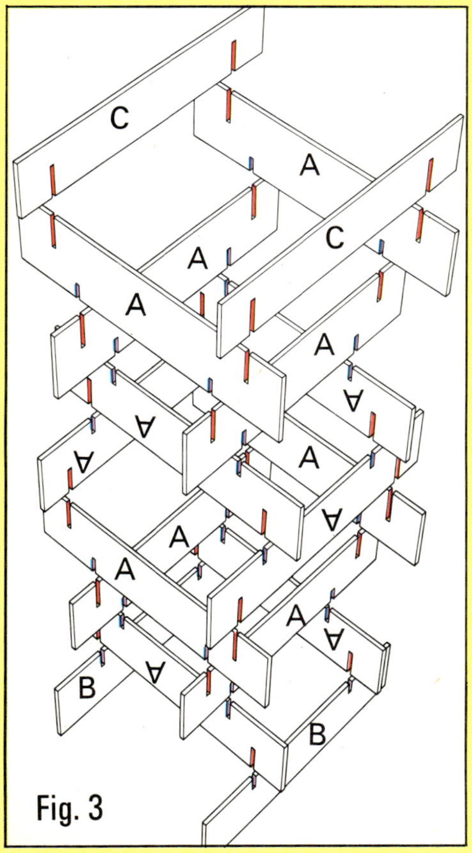

Fig.2 *(above). The cutting diagram for the slotted table lamp (parts shown in* red*) and the strung ceiling light (parts shown in* black*). This diagram is exactly half-size, so the $\frac{1}{2}$in. squares here will be 1in. across when you draw a full-size version.*

Fig.3. (left). An exploded view of the slotted table lamp. Most of the parts A have been omitted for clarity. Some of them fit in upside down; these are marked on this diagram with the letter A upside down as well.

Fig.4. (below, left). The flex for either of the table lamps can be passed through the side of the base and up through the middle by chiselling out a channel on the underside. This will not show when the base is upright.

Fig.5. (below, right). A more stylish version of the lamp base can be made of four pieces of wood arranged as shown, so that no end grain is visible. The pieces are simply glued together; when the glue is dry, a channel is cut for the flex as already described.

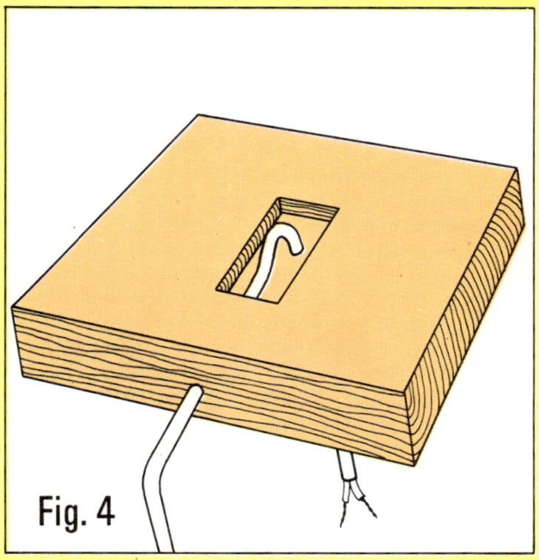

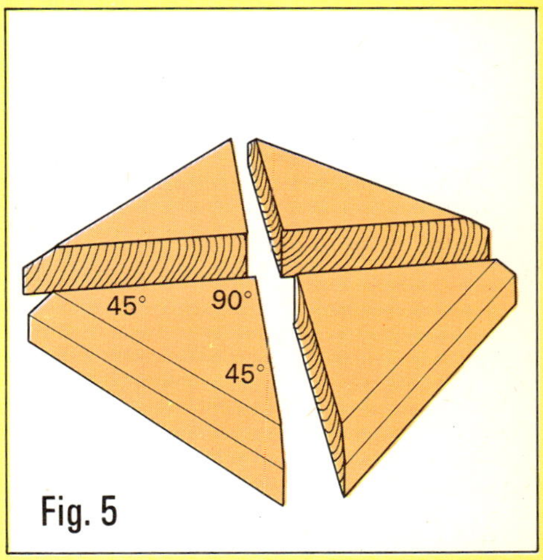

DGW

the protective paper to make the tape stick, but leave it on the outer faces of the two end pieces. Clean and polish the edges of all the pieces at once; this will save you a lot of time. Mark the position of the slots on the two end pieces.

Clamp the thirty pieces in a vice and cut all the slots to the correct depth with a fine, sharp tenon saw. Cut down the *inside* of the marked lines to avoid making the slots too wide. There will be a narrow 'tongue' of acrylic left inside each slot. Remove it with a fretsaw if you have one, or break it off each piece one at a time and clean up the bottom of the slot with the edge of a flat file.

Use the file to remove any rough places from inside the slots, then separate the thirty pieces and clean each piece with methylated spirit.

Cut the slots in the two pieces B and two pieces C by a similar method, but only two at a time instead of thirty.

Clean up the edge of piece D without removing the paper. Then mark the position of the centre hole for the light fitting and the eight small ventilation holes and drill them carefully as described above.

Assembly

The following instructions are for permanent assembly, but as it is easy to assemble the tower incorrectly—and not know it until the last two pieces do not fit—a trial assembly first is recommended.

Take two pieces A and place them on edge a short way apart and with the shallow slots uppermost. Then lower two more pieces A *the other way up* on to them at right angles, so that the two pairs of shallow slots engage and the four pieces form a hollow square. The exploded view, Fig.3, shows how the pieces fit together, but most of the pieces A have been omitted for clarity.

Chloroform the four pieces together using a brush in the ordinary way. Measure the diagonals of the square to make sure it is straight. Let the joints dry, then lower two more pieces A shallow slots down, on to the deep slots on the top edge of the two top pieces. Chloroform them in place. Continue to build the 'tower' of pieces A in this way until you have used all the pieces. The diagonals should be checked constantly, and one pair of pieces should be allowed to dry in place before putting another pair on it.

Complete the 'tower' by glueing the two pieces B (the ones with shallow slots) to the bottom and the two pieces C (with deep slots) to the top. This will finish the top and bottom off level. Leave the pieces to dry and fasten the 'batten' light fitting to the exact centre of the wooden base. Drop piece D over the fitting's threaded shaft and screw the retaining ring down on to it to hold it steady.

Lower the 'tower' over piece D and on to the base. Then use a wax crayon to mark on the outside of the tower the level to which piece D reaches inside. Lift the 'tower' off, take piece D off the light fitting and chloroform it in place at the level of the marks. Make sure it is straight and level.

Reassemble the light fitting and put a bulb in it to complete it. It looks best with a 100-watt clear bulb, if you can still get clear bulbs in your part of the world. You can also use this lamp upside-down and without a base as a hanging ceiling light. In either position, it will sparkle more if a piece of crumpled silver foil is placed in the tower, on the opposite side of piece D from the bulb.

The wooden base

At its simplest, the wooden base for these lamps can be made from a 9in. x 9in. x 1in. (229 x 229 x 25mm) block of the wood of your choice. Cut it dead square, finish the edges neatly and drill a hole down through the exact centre to take the flex from the light fitting.

The flex is made to emerge from the edge of the block by a simple trick. Drill a hole (of the correct size to take the flex) through the edge of the block as deep as you can make it. Then extend this hole to the middle of the block by cutting an open channel in the bottom of the block with a chisel. The flex can then be passed through the hole in the edge of the block, along

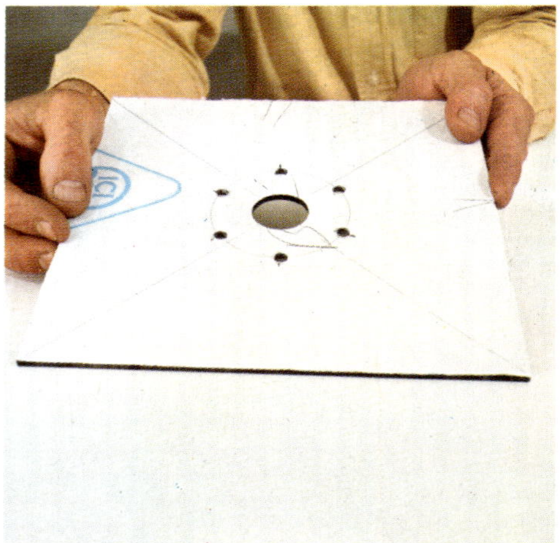
6

7

8

9

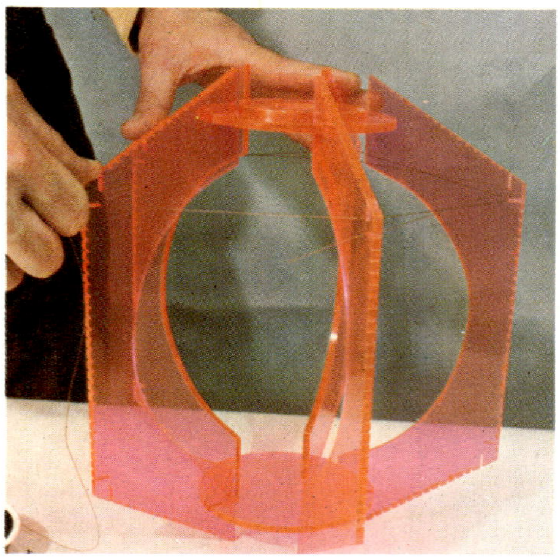
10

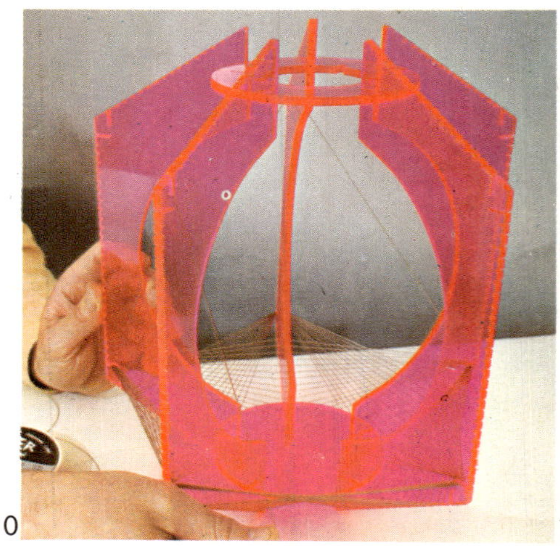
11

***6** (above, right). The protective paper is left on the acrylic while it is cut and drilled. This piece is the base of the box lamp.*
***7.** The pieces of the slotted lamp should fit together easily; do not force them.*
***8.** Fit pieces X and Y together to ensure that you glue the pieces W in the right place.*
***9.** The main structure of the ceiling light being assembled. Let it dry before stringing.*
***10.** The ceiling light being strung. Start at an end, where there are fewer slots.*
***11.** The sides are strung in the same way as the top and bottom. Start from either end; it makes no difference to the result.*

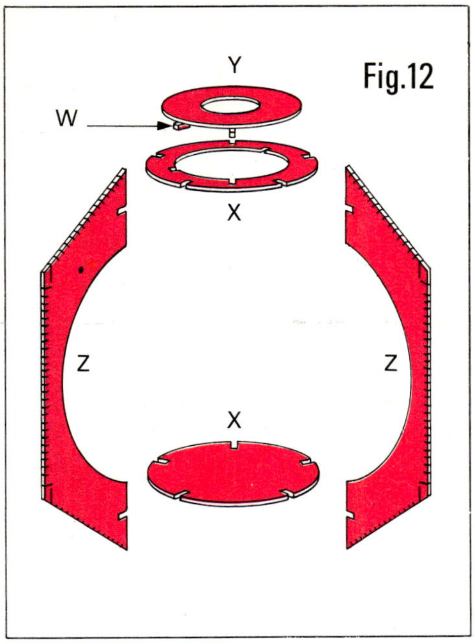

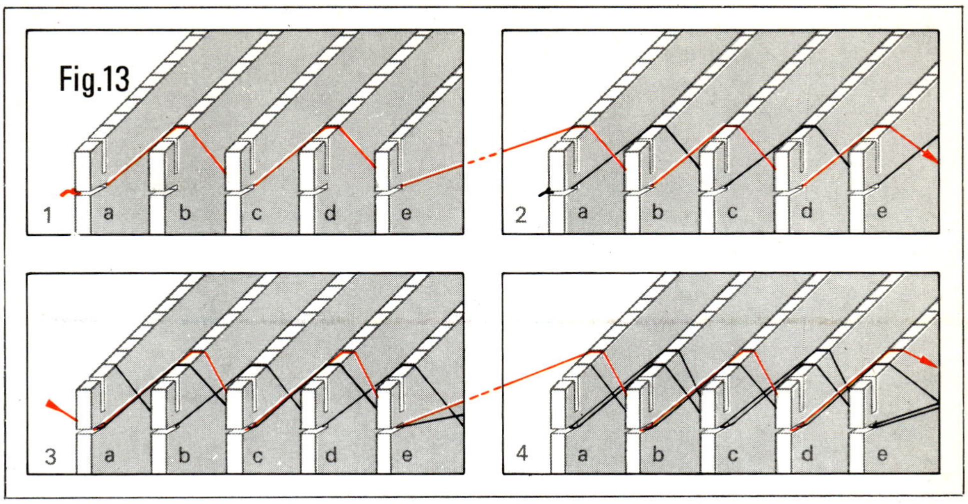

Fig.12 (left). An exploded view of the main frame of the strung light, showing how the removeable panel (piece Y) fits into the top to allow you to replace the bulb.

Fig.13 (above). A step-by-step stringing diagram. The circle of five pieces is shown as a line for clarity. The row being strung is in red; rows already done are in black.

the channel and up through the centre hole to the light fitting (see Fig.4). You can conceal the channel by sticking felt to the underside of the block.

A more elegant base is shown in Fig.5. It is made out of four pieces, so that no end grain shows. These pieces are in the shape of right-angled equilateral triangles (i.e., with one angle of 90° and two of 45°) with a base 9in. (229mm) long. You will need a mitre box to cut the angles (from a D.I.Y. stockist). Then just glue the four pieces together and insert the flex by the method already described, but wait till the glue is absolutely dry before you start.

The strung ceiling light

The ceiling light is made out of $\frac{3}{16}$in. (5mm) acrylic, slightly thicker than the other two. Its construction, though basically simple, is slightly complicated by the removeable panel on top that allows the bulb to be inserted and removed. This is necessary because the light is completely enclosed when all the threads are in place.

You will need a piece of acrylic 2ft x 1ft 6in. x $\frac{3}{16}$in. (610 x 457 x 5mm), an ordinary suspended bulbholder as used for most ceiling lamps, and 150 yards or metres of clear nylon fishing line – not too thick. You could use white button thread instead of the fishing line to give the light a different look.

The shape of the pieces is shown in *black* in the cutting diagram, Fig.2. Five pieces Z should be cut (do not cut the slots yet) and two pieces X. One piece X should be left solid; the other should have its centre cut out as shown. Cut also one piece Y and three of the very small L-shaped pieces W.

The way the pieces fit together is shown in the exploded view, Fig.12, and because of the rather complicated shape of the light you should study this diagram before going any further.

Stick the five pieces Z together with double-sided tape, removing all the protective paper except from the outer sides of the two outer pieces. Clean up the edges of all five pieces together, then put them in a vice and cut the slots exactly as shown. Note that each piece has two big slots $\frac{3}{16}$in. (5mm) wide, where other acrylic pieces fit, four slots $\frac{1}{2}$in. (13mm) deep and the same width as your tenon saw blade, where many strands of fishing line are to go, and 53 very small slots, where only one or two strands of line will fit. It is essential that these tiny slots should be absolutely equally spaced.

Stick the two circular pieces X together, clean up their outer edges and cut the five outer slots. Then separate them and cut a circle out of the middle of *one* of them as shown. This must be done very slowly and carefully with a fretsaw. Cut the three slots out of the inner circle. Note that they are not equally spaced, but are arranged to fall halfway between pairs of outer slots. Clean up the inner circle and the slots.

Cut out and clean up piece Y. The centre hole should be the right size to take the bulb-holder (normally $1\frac{1}{8}$in. or 28mm). Do not cut slots in this piece; the marks on the edge show where the pieces W will be glued on.

Cut out and clean up the three pieces W, which are little hooks to hold piece Y on to piece X. Tape the piece X with the hole in it on top of piece Y. Make sure it is dead in the middle, then stick a sharp spike through each of the three irregularly-spaced slots around its central hole and mark the position of each slot on to piece Y. Separate the two pieces and chloroform the three pieces W on to the marks on piece Y, with each 'hook' pointing outwards and its tip level with the circle's outer rim. You will then be able to fit piece Y on to piece X by passing the hooks through the slots, and lock it in position by rotating it a short way.

Assemble the rest of the shade as shown in the exploded view. It is now time to string it; before you start, study the stringing diagram (Fig.13) and also the photograph of the completed light.

You will see that the thread passes alternately through a large slot and a small slot. This applies equally to the stringing of the top and bottom and of the sides. The same large slot is used many times, but each small slot is used only once on the top and bottom, and twice on the sides. The large slots for stringing the top are just around the outer corner of each piece Z, on the sides of the shade. The large slots for stringing the sides are just around the corner on the top and bottom.

Bearing this in mind, start by stringing the top. Tie a knot in the beginning of your spool of thread or line, and insert it in the large slot at the top of the side of any of the pieces Z – *not* the large slot in the top. This piece is marked *a* in the diagram. Give the thread a tug to make sure the knot is big enough to hold it in the slot. Then carry the thread to the first small slot from the corner in the top of the next piece Z (marked *b*)—you can go clockwise or anticlockwise, as you prefer.

Next, carry the thread to the big slot around the corner in the third piece, marked *c* (i.e., the same slot as the first piece). Then to the first small slot in the top of the fourth piece, marked *d*. Then to the big slot in the fifth piece, marked *e*. Then to the first small slot in the first piece, marked *a*, and so on. When you have done two complete turns, all the first line of small slots will be filled, so for the next two turns, run the thread from the big slots to the *second* line of small slots, then for the next two turns to the *third* line, and so on for twenty-eight turns, by which time you will have filled all fourteen slots in the top of each piece. End by tying another knot in the thread and pulling it over the far side of the last slot – this is easy with nylon fishing line because it stretches like elastic.

Repeat this procedure to string the bottom. Each side is strung twice; once from the big slots in the top, and once from the big slots in the bottom. Perform each of these two stringing operations separately, using the same technique as for the top and bottom. The only difference is that during the second operation, there will be one thread in each small slot already. Ignore them and lay the second set of threads directly on top.

When the stringing is completed, finish the light by mounting the bulbholder (and a bulb – any size) on piece Y and slotting it into the top of the strung shade. It can then be hung up like any normal lampshade.

Decorate with candles

Over the past few years, the growing awareness of the decorative possibilities of candles scattered around various rooms of the home has resulted in a blossoming of the candlemaker's art. So many types and variations have been thought up and put onto the market that it would nowadays be virtually possible to burn a different candle every week of the year and still not run out of new ideas! With a minimum of practice you can easily make any number of quite diverse examples of this simple but satisfying art-form: moulded, carved, dipped, plaited, perfumed . . . No longer is a candle just that stub of white wax you rifle around for in the kitchen drawers every time the lights fuse.

Not only can you produce your own candles at a cost that compares favourably to the ready-made, store-bought article's, but with a little extra experience you can turn your own ideas into waxen reality. Isn't there something with a touch of the wry, the grotesque, or the off-beat you'd like to see burning away on the bed-side table?

Types of candles

Several examples of the types of candles you can readily learn to make are pictured here. Moulded candles fall into two broad categories—those which are cast in elaborate ready-made moulds, such as the latex rubber, metal, or glass shaping moulds available from craft shops or candlemakers' suppliers, and those that are cast in everyday household containers, such as cardboard boxes, plastic drainpipe sections, rubber balls and even glassware. These are often enhanced with variegated colouring, using the stearin-and-dye colouring mixture described below, so that while the basic shape of the candle is a simple one—round, spherical or cuboid of some description—there are layers, either diagonal, horizontal or vertical, of colour to give it that extra touch. Further refinements in these multicolouring techniques, such as 'chunky' colouring and the 'marbled' effect, are described below. A moulding liquid which enables you to take your own castings off an object you would like to have as a candle shape is also available to increase the scope of these moulded candles.

Sheets of beeswax, with their honeycombed texture and pleasant aroma of honey, can be rolled into candles, or of course the beeswax can be melted down and used as any other sort of candlemaking wax. By adding a little of this sort of wax to the more conventional paraffin wax you obtain a better shine on the outside surface of the candle, so it is well worthwhile buying some beeswax even if you are not sure whether you want to produce purely beeswax candles.

One of the oldest methods of candlemaking is 'dipping': pear-shaped and plaited candles are two examples of the possibilities of the dipping process. Dipping is especially suitable for making a so-called 'carved' candle, with layers of differently coloured wax set one after the other around the wick and then exposed to view when the candle is cut along its length like that shown on page 80. The rough-hewn look can also be obtained with this process by simply applying a little manual pressure to sections of the candle between 'dips'.

The final type of candle treated in this chapter is the 'sand' candle, which is moulded in a bed of damp sand carrying the impression of an ashtray, a bowl, an ornament—as in the case of the home-made moulds described above. As well as providing a mould, however, the sand also forms a hard crust around the outside of the candle which can then be worked into a design. Examples of this sort are shown on both the opposite page and overleaf.

Materials and requirements

Except for a little flair and imagination, all you will need to make any of the candles dealt with in this chapter may be obtained either from a candlemakers' supplier or a craft shop. The primary requirement, wax, can be either beeswax or paraffin wax, which is cheaper and therefore more widely used for most types of candles involving appreciable quantities of wax. The grades of paraffin wax available vary through a range of melting points—the best for the do-it-yourself candlemaker is the fully refined paraffin wax melting at between 135° and 140°F (57° and 60°C). Paraffin wax comes in both block and powdered forms, but it is recommended you use the latter. Coloured and plain sheets of beeswax, which can either be melted down and cast, or rolled into the sort of candle shown on the next page, are available as well as the customary block form.

Candlewick lengths are woven from bleached linen thread, and are generally sold in packs graded in terms of the diameter of wax they will burn. For example, a 'one inch wick' will burn down a cylinder of half-inch radius through the length of the candle. It is obviously crucially

Above right. *Three dipped candles, worked up into a plait. This unusual but simple and visually effective design makes for a candle capable of throwing off a good deal of light.*
Right. *Two moulded 'marbled' candles.*

important that you use the appropriate wick grade for the size of candle it is to burn, as too small a wick (i.e. where the area burnt is only a small fraction of the total area) will result in the wick being drowned in a pool of molten wax.

Since it is the particular colouring you give the candle that determines the greater part of its character, you should be careful where the wax dyes to be mixed in are concerned. The dyes themselves also come in both the powdered and solid forms, and once again, as only a comparatively small amount is required, it is the powdered form you should prefer. The dye is first dissolved in stearin, which is a flaky white species of wax promoting uniform colour suspension in the final product. To give you an idea of the quantity of stearin you should buy, allow one part of stearin to every ten parts of wax.

You should also have a sugar thermometer, reading up to a temperature of 400°F (205°C) While there is not much danger of your overheating the melting wax to its flash point (i.e. to the temperature at which it bursts into flame), the properties of wax change with its temperature—so to attain the exact effect desired, a reliable means of measuring its temperature is a necessity. Should you inadvertently heat the wax beyond its flash point, turn off the stove and try to cover the melting pot—do not attempt to carry the pot to the sink and be extremely careful if you have to douse it with water, as there is a danger of splashing the hot wax onto yourself. Take extra special care and check that fire precautions are adequate.

If you are using ready-made candle moulds, you will also need a mould sealing agent such as plasticine or clay, a wicking needle to run the end of the wick through the mould top, and some pieces of timber to support the mould whilst the wax is setting (see Fig.3). Sand candles require a box of sand, deep enough to take the object used to make the mould in the damp sand and still leave at least one inch of sand.

As far as utensils are concerned, you will need two large saucepans—one for melting the wax and the other for dissolving the dye-stearin suspension—and an enamel jug with a pouring lip for the molten wax. Dipped candles are best made with a deep enamel jug, which obviously must have a neck that is not too narrow. Finally, a large receptacle to act as the cold water bath (used to cool wax quickly) is required, as well as some spoons for ladling the hot wax from the melting pot into the moulds.

Preparing coloured wax

While slowly melting the paraffin wax in the first saucepan, take one part of stearin for every ten parts of paraffin wax to be dyed and melt it in the second saucepan. Then add the dye to the

Above left. *Some of the candles dealt with here (from the right): a carved candle, rolled beeswax sheet candles, a ball-moulded candle, chunky candles, two-colour candles, pairs of dyed dipped candles (suspended), a candle with its surface worked into a relief pattern with a knife, and a plain moulded candle bearing a flower motif.*
Left. *More home-made candles, including the twisted, coloured, moulded and sand varieties.*

stearin melt, and stir it until all the particles of dye are completely dissolved: the stearin-dye mixture can then be added to the molten paraffin. If you are adding a candle perfume, follow the instructions of the particular perfume manufacturer as regards the best time to include the perfume. As many perfumes are rather volatile, too high a melt temperature may mean that most of the perfume's aroma is vapourized off before the candle is finished.

It is recommended that you run a preliminary test on the strength of the dye you are going to use, since too much dye will reduce the glow given off by the candle. Take a spoonful of the molten wax you have coloured with the dye and drop it into cold water. As it congeals you will be able to get a good idea of the colouration imparted by that strength of dye—for the general run of colours, a small pinch of powdered dye will suffice for a pint of wax.

One very simple way of 'making' a coloured candle is to dip a plain white one in molten coloured (and perfumed, if you wish) wax. Float a two inch layer of molten coloured wax on water heated to a temperature of 180°F or 82°C, and dip the candle through the wax and into the water. As the candle is drawn out, a coating of the coloured wax will adhere to the surface of the candle. You must use the thermometer here to keep the water at the right temperature, as wax that is too hot will not allow enough colouring to be picked up, while a flaky outer covering will result if the wax is too cool. The dipping procedure can obviously be repeated either for increased depth of the coloured layer—don't overdo this, as the wick in the original candle will probably be graded only to burn a diameter of the order of that of the original candle—or for achieving a harlequin-like effect along the length of the candle with several different colours of wax. This is done by successively decreasing the depth to which the candle is immersed in the various colours used, allowing sufficient time between dippings for each layer to set properly.

Dipped candles

All you need to employ this time-honoured method, besides the wax and wick, is an enamel jug that is a little deeper than the required length of candle, a bath (e.g., a saucepan) of cold water, and a sugar thermometer. One end of the wick length is tied to a small stick—a pencil will do—and dipped into the jug which has been filled with wax at 180°F or 82°C. Withdraw the waxed thread and allow the coating to harden, which should take something like thirty seconds. Repeat the process until the desired thickness has been layered onto the candle, but watch out for the cooling of the wax in the jug, which is inevitable. Once the tell-tale bubbles that indicate cooling have started to appear on the surface of the wax, or on the last layer put onto the candle, reheat the wax to its original temperature and recommence dipping. When you have finished the candle, hang it up to let it harden naturally. Where the candle is the same colour throughout, an extra shine can be given to its surface by a final dip in wax heated to 200°F or 93°C, followed quickly by plunging it into a cold water bath.

A measure of variation can be introduced into the dipping method in three ways: by hand-moulding the candle between individual dips, by using several colours of wax for successive layers, and by plaiting two or three dipped candles of differing colours while they are still soft. As an example of the first variation, a pear-shaped candle can be made by taking one that has been dipped to a diameter of around half an inch and then dipping it to only half way up its length, or pouring wax over its bottom half. As the lower part of the candle thickens out, a little pressure applied between dips will serve to mould it into the shape you have in mind for it.

A so-called 'carved' candle is made by using the dipping process to build up layers of different colours around the wick—each layer should be about a quarter of an inch thick, and of course you can work the wax between dips into a shape a little out of the ordinary. The hardened candle is then cut in some fashion along part of its length, like those shown on page 80 and this page, so that the different colours at each level are exposed. In this case, the wax must be fairly strongly dyed, as otherwise the different colours will show through each other on the finished product and so spoil the effect. The carved candle may be dipped in wax floated on hot water, as described above, but you must make a central core of at least 1in. (25mm) by the usual dipping process to prevent any water getting onto the wick.

To twist a dipped candle, take one that is still soft, lay it on a clean smooth surface, and flatten it gently with a rolling pin. Square off the base and turn one end one way and the other in the opposite direction at the same time to produce the twist along the entire length of the candle. A plait of two, or preferably three, dipped candles can be made by having someone hold the candles by the base and when they are, again, still soft and workable, plaiting them together. If the right colours for the three candles are selected the final result can be most pleasing—and very illuminating!

Moulded candles

Ready-made candle moulds in either latex rubber or metal are available in a wide variety of shapes from craft or candlemakers' suppliers shops—the latex ones in particular should be treated with some care, and should be washed and dried after use. Latex moulds are generally not provided with wick holes, so you will have to use a wick needle to make one in each of the moulds you are using. Fig.3 shows a latex mould racked up, with the wick pulled fairly taut through the mould from its supporting piece of wood to the wick hole, which is blocked up with a sealing agent, such as plasticine, once the wick has been led through the hole. The same wick hole should be used every time the mould is.

With the mould securely racked up, pour the wax (heated to 180°F or 82°C) slowly into the mould, tapping the sides gently to get the air bubbles to the surface. Soon after it is poured, the wax will form a small well around the wick as it contracts: you should prod around the surface and, once it has been broken up, top up the mould with wax heated to the same temperature as before. Repeat this procedure

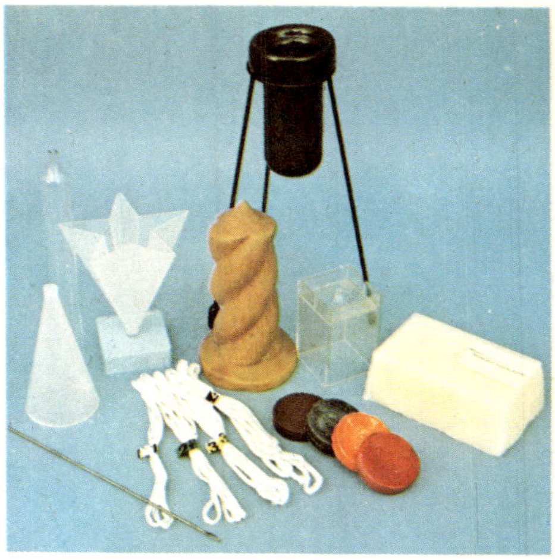

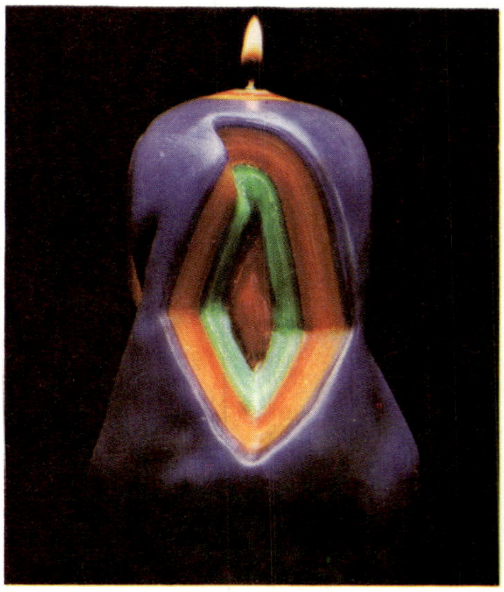

Top. *Candlemaker's equipment: candle moulds in a variety of shapes and materials, mould stands, graded wick lengths, a steel wicking needle, dye discs, and block beeswax.*
Middle. *Imaginatively worked 'sand' candles.*
Bottom. *An example of the 'carved' candle.*
Fig.1 (top right). *Building up candles using the dipping process. Two identical candles are often made at the same time, using both halves of one continuous wick length.*
Fig.2 (right). *A honeycombed sheet of plain or coloured beeswax can be cut down as shown and simply rolled into a fragrant candle.*
Fig.3 (right). *Moulds can either be racked up with a purpose-made wire stand, or, as shown here, on a makeshift timber support.*
Fig.4 (right). *Diagonal layers of coloured wax may be quickly set in a water bath.*

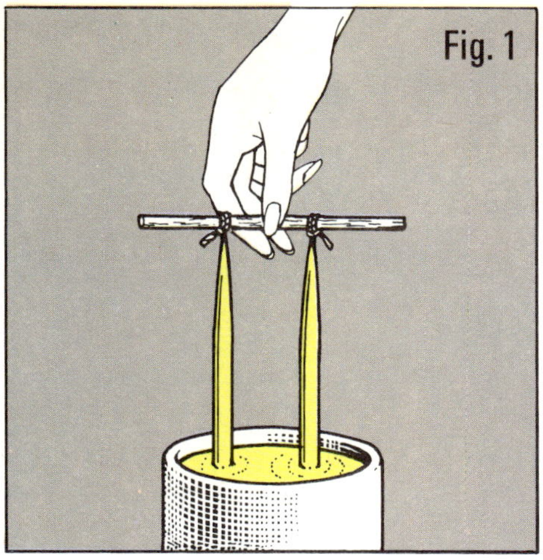

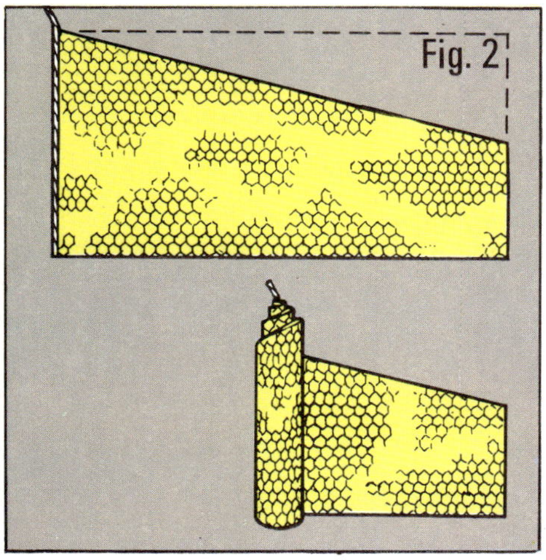

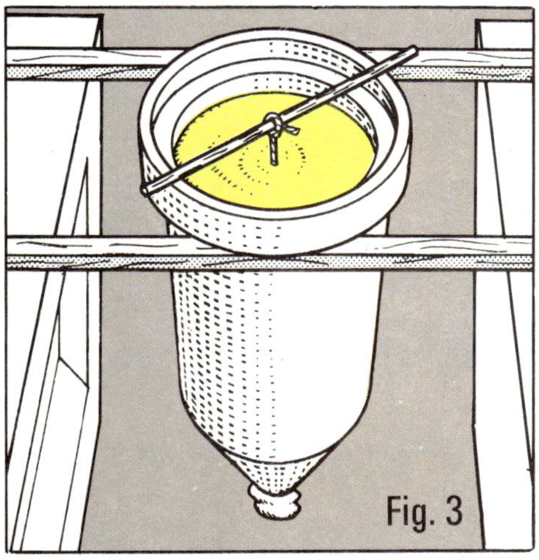

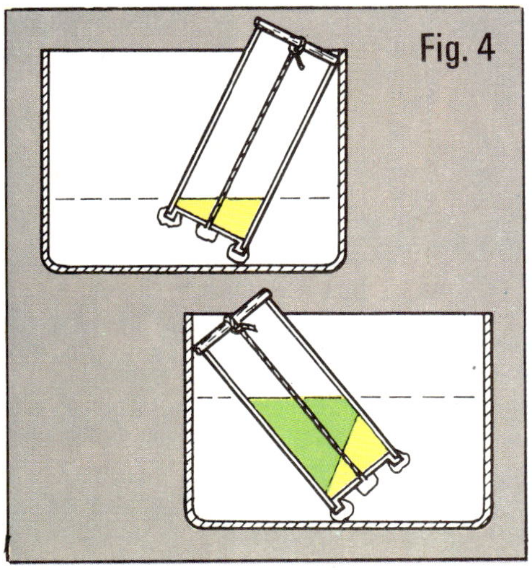

until the candle's base remains smooth upon cooling. Any minor aberrations of shape can be worked on with the hands while the candle is still hardening inside the mould. When the candle has cooled down, and is quite hard, wet your hands with soapy water and work on the mould until it can be peeled back. Return the mould to its normal shape, and remember to wash and dry it before putting it away.

The outer surface of the candle can either be polished with your hands or with a soft cloth—or, as mentioned above, a little beeswax can be added to the paraffin wax while it is being cast to give the finished product an attractive shine.

If you are making your own mould, using the moulding liquid mentioned above, to take cast a candle shape from something like a statue or a solid geometrical figure, ensure that the object used is free of dust and grease. The object should be gently heated in front of a fire and immersed in the moulding liquid: when you have taken it out and it has stopped dripping, hold it in front of a fan heater until the white of the setting liquid changes to yellow. This procedure is repeated—make sure that any air bubbles are blown off the surface, and do not touch the actual surface of the mould—until a uniform thickness of about $\frac{1}{16}$in. (1.5mm) has been built up. When this has completely set, trim the uneven edges off the mould and peel it off the object used as the former with the aid of some liquid soap. A wick hole can then be made at the top and the mould used in exactly the same way as a ready-made one.

Multi-colour effects

Candles incorporating a variety of colours are best cast in fairly large and simple moulds, such as sections of plastic pipe or cuboid household containers. 'Chunky' candles like those shown on page 80 are made by pressing pieces of coloured wax onto the sides of the mould and pouring wax at 200°F (93°C) over them to fill up the mould. This is then allowed to stand for one minute, and then it is placed in a cooling bath and the wicking needle run down the centre to form the course through which the wick will later be inserted. When the candle has set, remove the mould and candle from the bath and immerse it in water just under boiling point (212°F or 100°C is the boiling point of water). This serves to melt the outside of the candle and thus blend the various chunks into the body of the candle. Top up the mould with wax at 180°F (82°C), allowing the molten wax to cover the outside of the candle as well as making up for the well formed around the wick by the contraction that cooling entails. The mould is once again immersed in a cooling bath and allowed to harden. You may then remove the candle and insert the wick.

A 'marbled' candle, like those shown on page 79, is made by pouring a wax-stearin mixture at 190°F (88°C) into a warmed, solid mould that is already fitted with a wick—the first stage is really the same as for a normal moulded candle. Let the wax stand for a minute in the mould to allow air bubbles to rise out of the body of the candle, and then immerse the mould and candle in a cooling bath for 40 seconds. Pour the unset wax back into the pan, so that only a wax shell will be left in the mould; then spoon a little pure dye into the mould and swirl it around rapidly and tip out the excess. Repeat this process with two or three different shades of dye until you have managed to put on some sort of pattern or design. Then pour in half a cup of candle wax at 180°F (82°C) and quickly swirl this around in the mould too, and tip off the excess into a separate container so that the main pot of white wax will not be tainted should the wax have picked up some dye as it was swirled around. Repeat this seven or eight times, to trap the colours in an envelope of white wax, and then fill up the mould in the usual fashion with plain white wax. For best results, keep the mould warm throughout the process and cut down as much as you can on any pauses between individual stages of the marbling.

Perhaps the simplest of the multi-coloured candles—and therefore the one that gives the amateur candlemaker the most scope for a little inventiveness—is one that uses layers (which may be vertical, horizontal, or diagonal) of different colours. Some examples of this type are shown on page 80, these are made by simply pouring layers of coloured wax at 185°F (85°C) into a mould, and allowing each layer to set partially, before introducing the next. For diagonal layers, the mould can be laid at an angle in a cooling bath and the same process used (see Fig.4).

Sand candles

Examples of this type are shown on page 80 a box of damp sand is used, with the object whose shape is to be cast pushed into a hole made in the middle of the box. The surface of the sand is smoothed, but do not pack down the sand in the box. Paraffin wax heated to 250°F (121°C) is poured into the centre of the shape made in the sand until the hole is filled. You will once again have to top up the hole a few times as the wax cools and contracts, and wax of the original temperature should be used for these additions. The hotter the wax is, the more sand will be picked up and stick to the sides of the candle—at this temperature, between $\frac{1}{2}$in. and 1in. (13mm and 25mm) will be coated on. If you are thinking in terms of a rather larger candle you can use wax at a higher temperature to add some bulk to the outside.

Allow the sand candle to set in a cool place for one or two hours, and then push the wicking needle down the centre and leave the candle overnight. The candle may then be dug out carefully, with the loose sand around the outside brushed away. Remove the needle and insert a wick which will leave about one inch of the wax unburned. Top up the candle with wax at 220°F (104°C) and allow it to set. The sand shell can be worked to lend the candle any sort of finished look that appeals to you, and if the right grade of wick has been used this shell will in fact remain intact and reusable.

These suggestions are of course no more than an introduction to the candlemaker's art—once you have attained competence and a 'feel' for the materials you are using, the door is open to you to produce some truly individual designs and shapes. In addition to its candlemaking applications, don't overlook the possibilities of wax as an easily moulded medium for purely ornamental castings.

All you need to know about roof lights

If you are converting your loft into an attic room, some form of natural light is essential. This can easily be provided for by a simple roof light or skylight, i.e. a window set in the roof. The addition of natural light to an ordinary loft area not only makes it a better proposition as a storage area, but turns it into a potential living room.

Roof lights have other uses apart from providing light and ventilation for lofts. If the position is suitable they can be fitted over a badly lit stairwell, or used to provide any room which is covered with a roof with additional natural light. If the roof is pitched and the room has a flat ceiling, the problem of placing the light can be overcome by creating an inverted well using plasterboard on a timber framework which extends from the ceiling level to the light set in the roof.

In Britain, certain regulations govern the installation of rooflights if the room they illuminate is to be used for living purposes. Not only must there be a minimum area of glass to floor area, but a certain part of the window must allow for natural ventilation. Mechanical ventilation, such as an extractor fan, can be used and may in fact be preferable if you want the rooflight to be watertight. In Britain the minimum area of the light which must have a ventilating function is 1/20th of the floor area, and some part of the ventilating area must be at least 1.75 metres above floor level. It is more difficult to specify exactly how large the window should be but as a general rule a minimum of 1/10th of the floor area is a fair standard.

For completeness, this article should be read in conjunction with the roofing chapters in HOME CARPENTER 35-36-37.

Planning permission

The possibility of requiring planning permission depends on a number of factors, such as the situation of the site, etc. The short answer is to make sure in all cases that plans of your proposals are submitted to your local authority planning department for planning and building permission — before work begins. HOME DESIGNER 31 details the correct procedure for applying for planning permission.

Simple roof lights

An example of a simple roof light is shown in Fig.3. There are two main component parts apart from the glass—the curb on which the window frame rests, and the frame itself.

The curb is house jointed and screwed together, as shown in Fig.3, and can be fitted

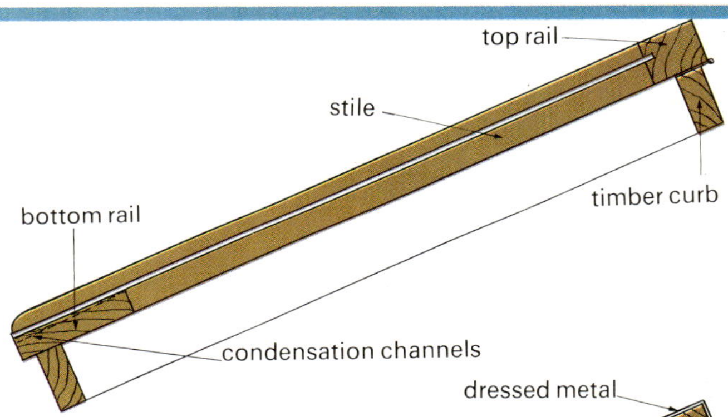

Fig. 1

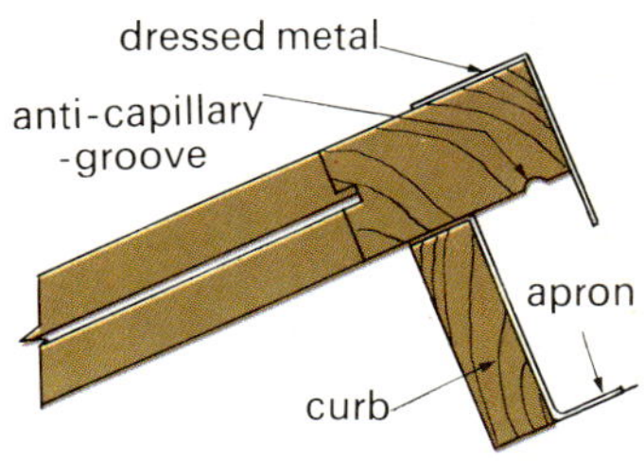

Fig. 2

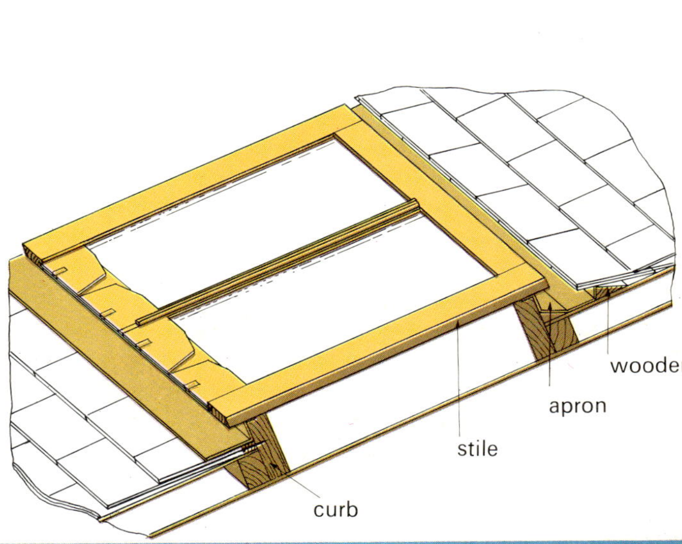

Fig. 3

Left. *A typical roof light on a pitched roof. This type provides light only.*
Above right. *Roof lights over living areas make them light and airy.*
Fig.1. *Side elevation of hinged roof light which shows how the watertight top rail differs from the bottom rail.*
Fig.2. *Side elevation of the top rail and curb. A metal shield is dressed over the top of unhinged roof lights to keep out rain.*
Fig.3. *A waterproof apron of lead, zinc or roofing felt surrounds the roof light.*

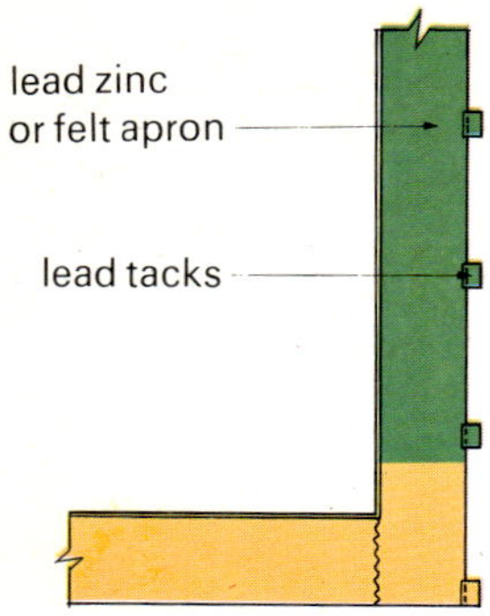

Fig. 4

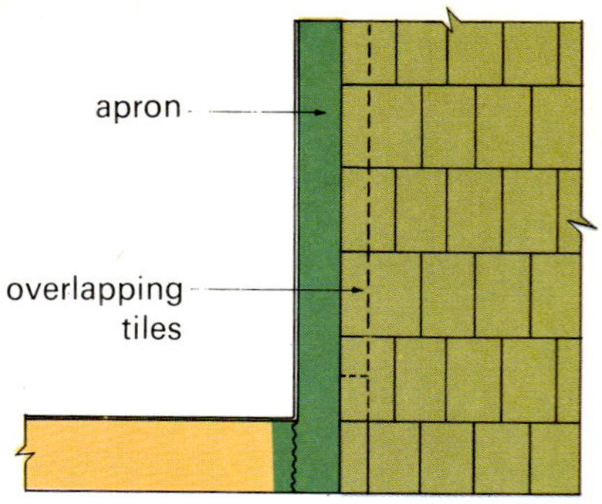

Fig. 5

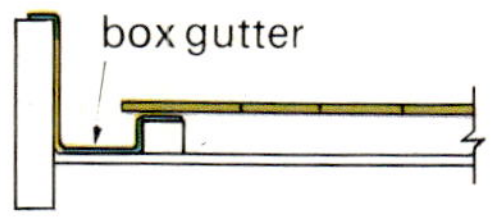

Fig. 6

Fig.4. *A lead or zinc apron is fastened down with lead tacks, hammered in round the edges of the apron without piercing it.*
Fig.5. *The slates above the top rail overlap the waterproof apron.*
Fig.6. *The waterproofing material is dressed up over fillets to form a simple box gutter.*
Fig.7. *When the space for the roof light has been cut, the adjacent rafters are strengthened and the sides trimmed.*
Fig.8. *A more elaborate roof light for a pitched roof is this raised type, which provides more storage space.*
Fig.9. *A lantern light provides the maximum amount of storage space.*
Fig.10. *Side elevation of one corner of a lantern light.*
Fig.11. *Plan of one corner of a lantern light. The sills are mitred together at the corners and are joined with tongued and grooved joints. The corner post is box and tenon jointed into the sill.*
Fig.12. *The mitred and jointed sill.*
Fig.13. *End of boxed tenoned corner post.*
Fig.14. *Plan view of two head pieces tenon jointed into a corner post.*
Fig.15. *The pitched roof rails joint together to form a ridge which is capped by lead or other waterproof material.*
Fig.16. *An alternative method is to fix a ridgeboard in which the roof rails are housed.*
Fig.17. *A plan view of the side and front of the raised roof light shown in Figs.8 and 18. These pieces are joined with tongued and grooved joints.*
Fig.18. *An exploded diagram of one end of the raised roof light shown in Fig.8. The top, which is pitched at an angle of 10°, fits over the side and front frames which are fixed together by means of tongued and grooved joints as shown in Fig.17.*

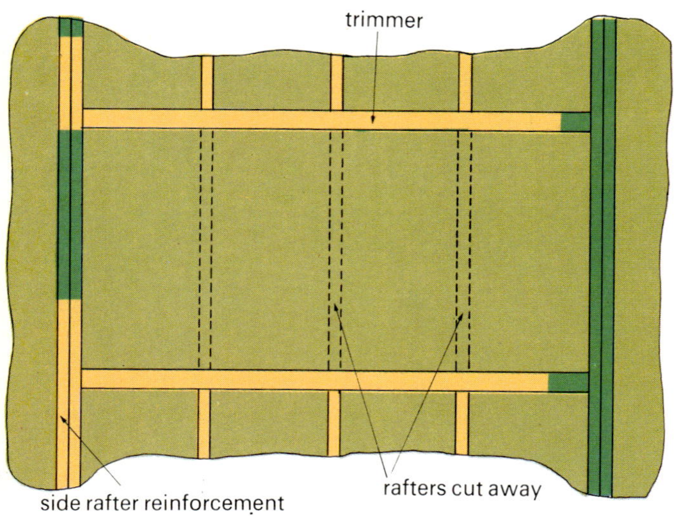

Fig. 7

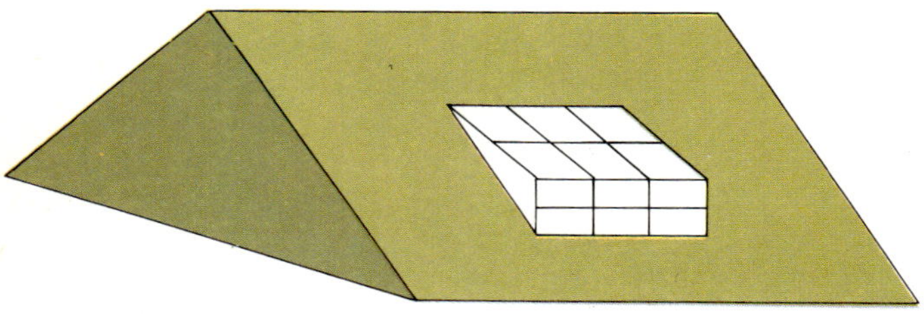

Fig. 8

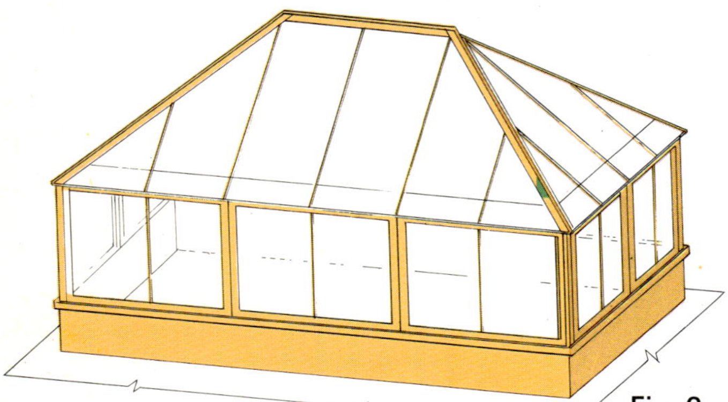

Fig. 9

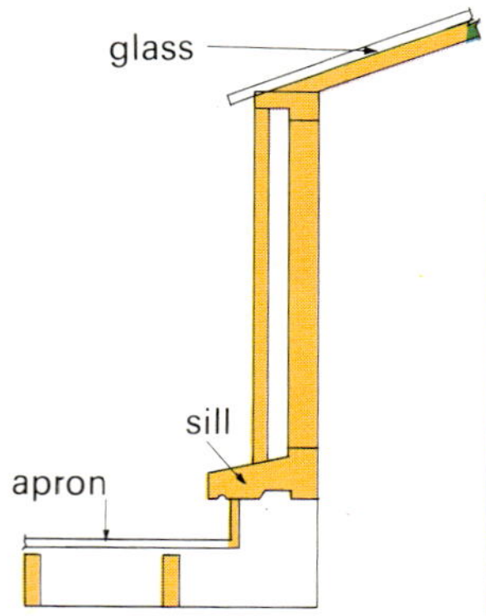

Fig. 10

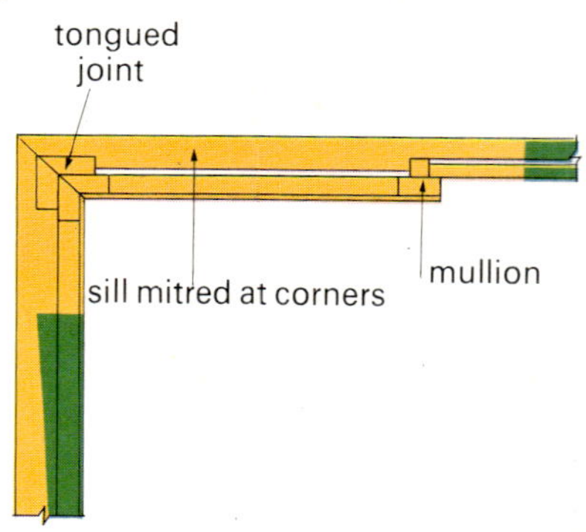

Fig. 11

Fig. 12

Fig. 13

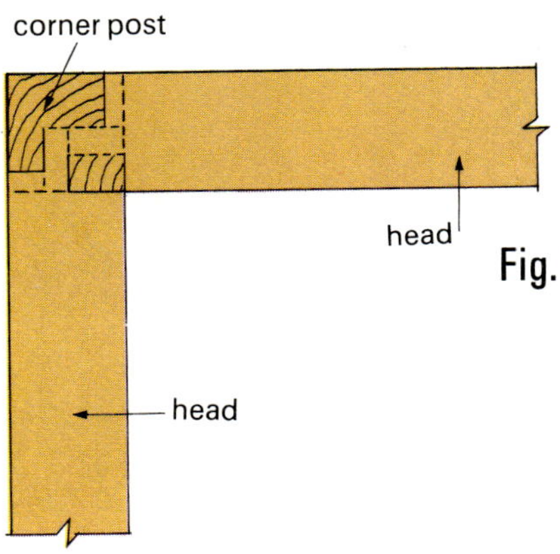

Fig. 14

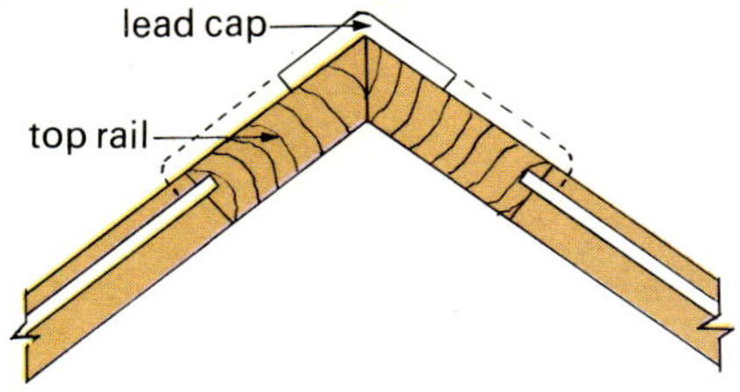

Fig. 15

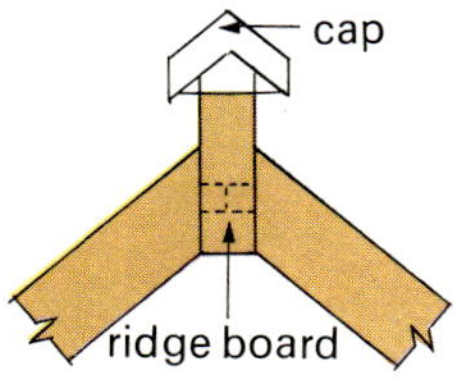

Fig. 16

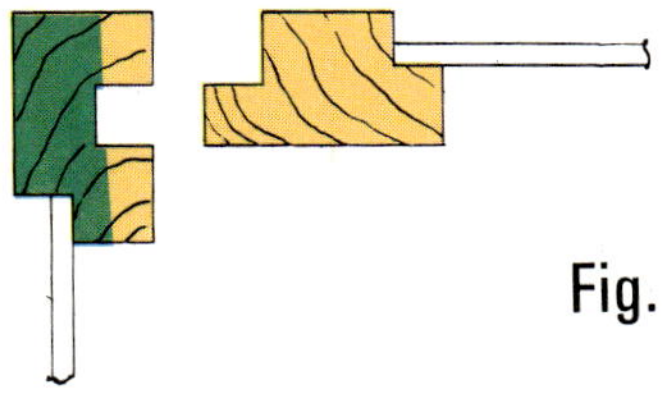

Fig. 17

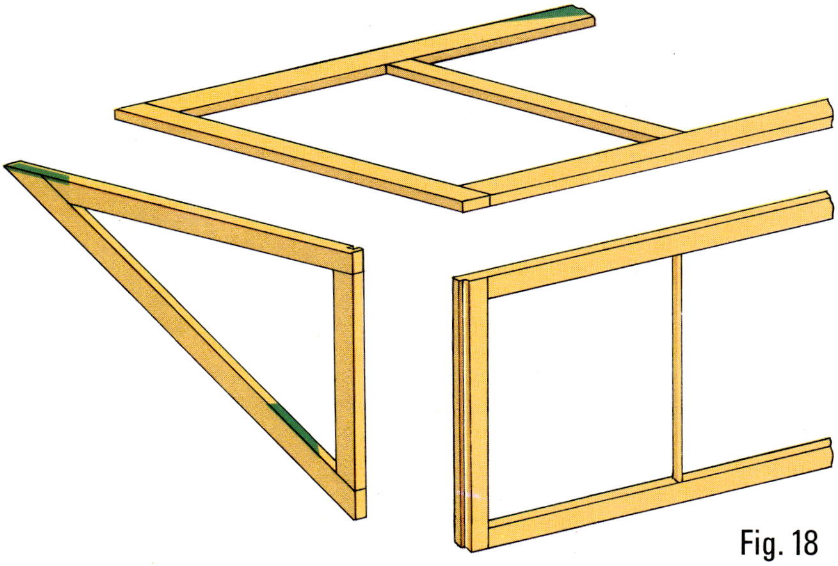

Fig. 18

either to the top of the roof rafters, using skew-nailing or simple brackets, or it can be fixed between the rafters.

The frame can be assembled using mortice and tenon joints or halving joints. To take the glass, the top rail is grooved and the stiles and any glazing bars are rebated. The bottom rail is thinner than the top rail and the stiles and the glass lies over the top of it. The reason for this is to allow condensation, which forms on the underside of the glass to run down the window and out onto the roof surface through narrow grooves or channels cut into the surface of the bottom rail.

The roof light may be permanently fixed or hinged depending on whether it is going to provide light only or ventilation and/or roof access. If the light is to be hinged the hinges must be fitted to the top rail. Use brass or galvanised metal hinges and ensure that the hinged edge will be watertight. This is particularly important if the roof light is to open wide for access to the roof. Fig.2 shows the features which prevent water getting through the light. A sizeable overhang on the top rail is provided together with an anti-capillary groove which runs the length of the top rail. This groove can be cut with a routing plane, but a V-shaped cut made with a handyman's knife would suffice. All but driving rain would be excluded by this method.

If only ventilation is required, the roof will not need to open completely and the addition of a metal 'shield' to the top rail, as shown in Fig.2, will be an added safeguard against rain penetration. This can be made of galvanised sheet metal which is dressed over the edge of the frame. Hinged metal rooflights can be purchased, but while these afford better weather protection because they overlap the curb, they do tend to be more prone to condensation. If you are purchasing a metal roof light make sure that the dimensions of the unit will be suitable for the proposed job. They come in standard sizes so you could have trouble with awkwardly placed rafters.

For safety's sake only heavy gauge or wired glass should be fitted to the frame. If you have any doubt consult a manual of glazing techniques and the various types of glass.

A final warning word. Ensure that you use a strong catch and hinges to prevent the frame from lifting off during a gale.

Cutting the opening

The first job is to remove a section of the roof surface and roofing battens in order to get at the rafters where you intend installing the roof light. Use a slate ripper to remove the fixing nails from the slates or tiles, some tiles will probably not be nailed and can simply be lifted from the roof battens. When you have removed the slates or tiles they should be stored carefully until they are needed for relaying. To allow you a clear working space always make a larger opening in the roof surface than required for the actual roof light.

In most cases a number of rafter sections will have to be cut away to make the opening. The side rafters will have to be reinforced and the opening trimmed as shown in Fig.7. The full length rafters at the sides of the opening can

Above. *A large but windowless loft area has been successfully converted into a cheerful living room by the addition of a large single pane roof light.*

be reinforced in a similar way to openings in floors for staircase wells, ie. doubling up on the rafters. The side rafters added across the line of the existing rafters should run the full slope of the roof from wallplate to ridge head.

Factory made prefabricated roof trusses cannot be cut away like independent rafters as the whole roof structure would be weakened. Because of this there is little point in making a raised roof light for storage purposes because the trussed rafters would have to be left to intrude into the roof space. If your house has such a roof it is possible to create additional natural lighting by the use of special transparent tiles that are fitted in the same way as normal roof tiles. Check with your local authority before fitting these as there are restrictions on the number that can be used.

Weatherproofing

When the roof light is installed the edges must be properly weatherproofed. Lead, zinc or roofing felt can be used, the latter being perhaps the easiest material to use. Figs.4 and 6 show the various methods of weatherproofing the roof light. If roofing felt is used, the joints can be bonded with a bitumen compound or with hot pitch. If you are working in cold weather, ordinary roofing felt can be made more pliable by gently warming with a blow lamp.

Joints in lead flashings are made by lead burning, which is somewhat similar to welding a joint in steel. If you are using lead, note that some lap joints must be made to allow for expansion, which can be quite considerable with lead.

Raised roof lights

A rather more elaborate roof light for a pitched roof is shown in Fig.8. With this type of roof light it is important to ensure that rain falling on the top section will be able to drain away. To facilitate this, the top section should slope slightly downwards, a slope of about 10° being adequate for a roof pitch of 45°.

Construction details are given in Figs.17-18. The frame is constructed in four sections—two triangular end pieces and the top and front sections. This is the easiest way to avoid complicated three-way joints at the corner posts. Using this method, fairly light timber can be used, say 3in. x 1½in. (75mm x 38mm). The joints between the various sections are tongued and grooved, and on assembly, synthetic resin glue and galvanized or sheradized screws should be used. The roof light should be weather proofed as described above.

Roof lights in flat roofs

A house with a flat roof is unlikely to have a loft space, but a 'lantern' roof light can be fitted to provide extra light to a room, staircase etc. An example of a lantern light is shown in Fig.9. This type is quite suitable for older properties, but the style is rather outdated for modern ones. Instead, purpose-made perspex domes can be obtained to suit modern house styles, or you could alter the design shown, say by making the top section virtually flat. (Remember that a slight slope must be included to throw off rainwater).

The construction of the lantern light shown needs to be fairly sturdy. The ends of the lantern can be formed as a gable roof or as a hipped roof —in both cases shaped glass panels will have to be used for part of the construction.

Full construction details are given in Figs.10-16. Note that as in the case of the rooflight in a pitched roof, the sides of the openings will have to be strengthened. Trimmers will be required, and the rafters can be reinforced, in most cases, by bolting larger rafters to them, as shown.

ELIZABETH WHITING

Light is bright

There is no need to despair if your house or flat is dark. If you use some ingenuity in the way you decorate and furnish, you can transform the dull rooms in your home into bright, attractive areas.

Here are some decorative ideas for dark rooms. The point to remember is that furnishings and decorations must be co-ordinated. In order to achieve a total look you must also choose the right lighting and accessories.

Use these suggested colour schemes in the dark rooms in your house, and you will be surprised at the effects they create. Add these ideas to those you have already found in the previous chapters for a light, bright home.

Left. *A dark room will become brighter by using light decorations. The effect won't be cold if splashes of a warm colour are added.*

ROOM	WALLS AND CEILING	WOODWORK	FLOOR	WINDOWS	LIGHTING	COVERS AND CUSHIONS	FINAL TOUCHES
WHITE LIVING ROOMS, HALLS AND BEDROOMS	white paper or paint	white	white-painted boards or white rugs on sealed floor, or pale oatmeal, beige, or white carpet	plain white blinds or curtains; add a coloured border for extra effect, or use coloured blinds or curtains with a white border, for a feeling of warmth	white paper shades or white spotlights, white table lamps with white shades	either plain white or anything with a predominantly white pattern	fresh green plants in white pots, a big mirror in a white frame
WHITE KITCHENS	white paint or tiles	white skirtings etc and white unit fronts with natural wood frames; worktop in white or natural wood	cork tiles, white tiles or predominantly white sheet flooring	white pvc curtains or washable blinds —these can be holland or Venetian blinds	concealed strip lights over work surfaces, or downlighters, spots, or fluorescent strips	brightly coloured squabs on kitchen benches etc	white storage jars or boldly coloured accessories
WHITE BATHROOMS	white gloss paint or tiles	white or a bold colour like buttercup yellow, apple green or kingfisher blue	cork tiles or white sheet flooring or tiles with a big shaggy bath mat	bright patterned curtains or blinds in any colour scheme on plain white; white shutters or a plain white lacey blind	downlighters above basin and bath, or ordinary white spotlight, or a paper shade (good cheap alternative for a central light fitting you can't move), or opal glass light fitting	stool with white pvc seat, or seat covered to match the curtains or blind	plant in white pot, junk shop mirror with white-painted frame, gay plain or patterned towels
WHITE DINING ROOMS	white	white	sanded and sealed boards, or white-painted boards, rush matting or white tiles	plain white blinds or curtains with coloured border	rise and fall lamp over table or white spotlights or downlighters to highlight the corners of the room	these can either be white to keep up the white feel of the room or in a bold colour to match the china or border on the curtains	white table and chairs, white ornaments, white china

ROOM	WALLS AND CEILING	WOODWORK	FLOOR	WINDOWS	LIGHTING	COVERS AND CUSHIONS	FINAL TOUCHES
YELLOW LIVING ROOMS, HALLS AND BEDROOMS	yellow all over, or yellow walls and white ceiling, or yellow with white cornice, or gay yellow paper	white or yellow to match walls exactly	sanded or sealed boards with white rugs or yellow mats, or white or yellow-painted boards with white rugs; carpet in yellow, straw, gold, beige, white, grey or to tone with walls	yellow or white blind or curtains; choose patterned or plain fabric with or without a border	yellow table lamps with matching or white shades; white spots or downlighters to highlight a special feature or a super ornament in a dark corner	white or yellow to match or tone with the curtains; if curtains are patterned, choose plain covers and vice versa	bright pop posters and a yellow mobile in one corner
YELLOW KITCHENS	yellow walls and white ceiling, or white walls and yellow tiles and ceiling, or yellow patterned paper on walls with white ceiling	white with yellow unit fronts; work surface in white, yellow or natural wood grain	cork tiles, or yellow patterned sheet flooring, or yellow and white checkerboard tiles, or plain yellow or plain white tiles	yellow or white laminated blind; yellow or white curtains; yellow or white shutters	as for white kitchens	bright yellow or white with appliqué lemons and sunflowers	yellow pots and pans, white or yellow storage jars, yellow jugs and mugs on display
YELLOW BATHROOMS	yellow paint all over, or yellow wallpaper with white ceiling–choose the warmer shades like buttercup and daffodil yellow rather than lemon which could be too cold for a dark bathroom	white or yellow to match walls	cork tiles, or yellow sheet flooring, patterned or plain, or yellow tiles or yellow or white carpet	as for yellow kitchens, or patterned café curtains in a laminated fabric	as for white bathrooms	seat upholstered in yellow	a row of yellow plastic ducks, yellow towels to match walls, or white towels with a yellow edging to match walls (you could easily do this yourself with plain binding)
YELLOW DINING ROOMS	yellow walls, white ceiling or vice versa; all-over wallpaper; white or yellow walls with deep yellow frieze paper round skirtings and door frames	white	white or yellow painted floor boards, rush matting, or cork tiles, sanded and sealed floor boards or toning carpet	gay yellow striped blind or curtains; shutters made from louvres or figured hardboard, painted white or yellow; white heavy net curtains on white pole	as for white dining rooms	white or yellow, patterned or plain	a big bowl of oranges and lemons, yellow paper flowers in a large corner vase, mainly white posters
ORANGE LIVING ROOMS, HALLS OR BEDROOMS	white ceiling, orange walls or vice versa, or all white or orange paper on walls, or orange hessian or felt	white	as for yellow, but with orange mats or orange carpet	white blinds or curtains with orange borders or a large appliqué	as for white	white, with cushions in yellow, white and orange or pale orange or apricot	modern mirror with orange perspex frame, an old brass oil lamp, pop art posters or gay prints
ORANGE KITCHENS	white with orange and white tiles or orange wallpaper	white with orange unit fronts and white work surface	as for yellow, but choose orange instead of yellow shades	white or orange blind	as for white kitchens	kitchen benches and seats in white	copper and white storage jars, white clock
ORANGE BATHROOMS	white ceiling, orange walls	white	orange or white sheet flooring, tiles or carpet	orange patterned or plain blinds, white Venetian blind, or white shutters	as for white bathrooms	white	orange-framed mirror, orange tooth mugs and towels
ORANGE DINING ROOMS	white ceiling, orange walls, patterned or plain	white	as for orange living rooms	as for orange living rooms	white rise and fall lamp over table, pale orange coloured Tiffany shades, or spotlights trained on ceiling	as for living rooms	a huge bowl of oranges, orange flowers, orange china
NEUTRAL SHADES–BEIGE, OATMEAL, CREAM, PALEST GREY	Follow the rules for white, substituting the colour you choose for white						
PASTEL SHADES	Follow the rules for yellow, substituting the colour you choose for yellow						
BOLD COLOURS –APPLE GREEN, LIME GREEN, SCARLET AND BRIGHT BLUE	Follow the rules for orange, substituting the colour you choose, and use lots of white with it